Target
Get back on track

GRADE 3

Edexcel GCSE (9–1)
Mathematics
Algebra and Shape

Katherine Pate

Pearson

Published by Pearson Education Limited, 80 Strand, London, WC2R ORL.

www.pearsonschoolsandfecolleges.co.uk

Text © Pearson Education Limited 2017
Typeset by Tech-Set Ltd, Gateshead
Original illustrations © Pearson Education Ltd 2017

The right of Katherine Pate to be identified as author of this work has been asserted by her in accordance with the Copyright, Designs and Patents Act 1988.

First published 2017

19 18 17
10 9 8 7 6 5 4 3 2

British Library Cataloguing in Publication Data
A catalogue record for this book is available from the British Library

ISBN 978 0 435 18331 8

Printed in Italy by Lego S.p.A

Helping you to formulate grade predictions, apply interventions and track progress.

Any reference to indicative grades in the Pearson Target Workbooks and Pearson Progression Services is not to be used as an accurate indicator of how a student will be awarded a grade for their GCSE exams.

You have told us that mapping the Steps from the Pearson Progression Maps to indicative grades will make it simpler for you to accumulate the evidence to formulate your own grade predictions, apply any interventions and track student progress.

We're really excited about this work and its potential for helping teachers and students. It is, however, important to understand that this mapping is for guidance only to support teachers' own predictions of progress and is not an accurate predictor of grades.

Our Pearson Progression Scale is criterion referenced. If a student can perform a task or demonstrate a skill, we say they are working at a certain Step according to the criteria. Teachers can mark assessments and issue results with reference to these criteria which do not depend on the wider cohort in any given year. For GCSE exams however, all Awarding Organisations set the grade boundaries with reference to the strength of the cohort in any given year. For more information about how this works please visit: https://qualifications.pearson.com/en/support/support-topics/results-certification/understanding-marks-and-grades.html/Teacher

Each practice question features a Step icon which denotes the level of challenge aligned to the Pearson Progression Map and Scale.
To find out more about the Progression Scale for Maths and to see how it relates to indicative GCSE 9–1 grades go to www.pearsonschools.co.uk/ProgressionServices

Pearson has robust editorial processes, including answer and fact checks, to ensure the accuracy of the content in this publication, and every effort is made to ensure this publication is free of errors. We are, however, only human, and occasionally errors do occur. Pearson is not liable for any misunderstandings that arise as a result of errors in this publication, but it is our priority to ensure that the content is accurate. If you spot an error, please do contact us at resourcescorrections@pearson.com so we can make sure it is corrected.

Contents

Useful formulae

Unit 3 Graphs

$$\text{Speed } (s) = \frac{\text{distance } (d)}{\text{time } (t)}$$

$$\text{Average speed} = \frac{\text{total distance}}{\text{total time}}$$

Unit 5 Triangles

The angles in a triangle add up to 180°.

Angles on a straight line add up to 180°.

The exterior angle of a triangle is equal to the sum of the two opposite interior angles.

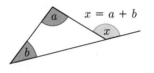

Unit 6 Quadrilaterals

Area of a rectangle = length × width

Area is measured in mm², cm² or m².

The angles in a quadrilateral add up to 360°.

Unit 7 Angles

Vertically opposite angles are equal.

Alternate angles are equal.

Corresponding angles are equal.

Glossary

Unit 1 Simplifying and brackets

Power: a number written above the base number that tells you how many times the base number is multiplied by itself, e.g. 5^4 is 5 to power 4 = $5 \times 5 \times 5 \times 5$.

Algebraic expression: a combination of letters, numbers and operations. The letters stand **for unknown numbers.**

Term: a number, a letter, a number and a letter multiplied together, or two or more letters multiplied together, e.g. 6, x, $3x$, $4ab$.

Like terms: terms that have exactly the same powers of the same letters, e.g. $4a$ and $3a$ or x^2 and $5x^2$.

Expand: to expand a bracket multiply each term inside the bracket by the term outside the bracket.

Factorising: the reverse of expanding. Divide each term by the highest common factor (HCF) of the terms and write the HCF outside a bracket.

Factor: a number or expression that divides exactly into another number or expression.

Highest common factor (HCF): if you find all the factors of two numbers, the highest number that is a factor of both numbers is the highest common factor (HCF).

Unit 2 Equations, expressions and formulae

Equation: a mathematical statement that contains an unknown number (a letter) and an equals sign.

Solving an equation: working out the value of the unknown number.

Inverse operation: the operation that 'undoes' an operation.

$+$ is the inverse of $-$, and vice versa

$\times$ is the inverse of $\div$, and vice versa

Variable: letters in algebraic expressions whose values can change or vary.

Formula: a general rule that shows a relationship between variables.

Substitute: put numbers in place of letters in an algebraic expression, equation or formula.

Unit 3 Graphs

Coordinates: values on the x- and y-axes on a graph.

x-axis: the horizontal axis on a graph.

y-axis: the vertical axis on a graph.

Distance–time graph: a graph showing distance on the vertical axis and time on the horizontal axis.

A horizontal line on a distance–time graph shows no change in distance, so the person or object is not moving.

You can use a distance–time graph to find the speed the person or object is travelling at.

Unit 4 Sequences

Sequence: a list of terms that follow a rule. In a number sequence the terms are numbers.

Index notation: a way of writing a number that is multiplied by itself. The number is called the base and the number written above the base is called the index or the power. The index or power tells you how many times the base must be multiplied by itself. For example, $5 \times 5 \times 5 \times 5$ written in index notation is 5^4; 5 is the base and 4 is the power.

Arithmetic sequence: a sequence where the terms increase or decrease by the same number each time.

Arithmetic progression: another name for an arithmetic sequence.

Term-to-term rule: the rule that you use to get from one term to the next in a sequence.

Unit 5 Triangles

Isosceles triangle: a triangle with two equal length sides and two equal angles at the bottom of those sides.

Equilateral triangle: a triangle with three equal angles and three equal length sides.

Line of symmetry: a line that divides a shape into two halves, where one half is a reflection of the other.

Parallel lines: lines that are always the same distance apart and never meet.

Interior angle: an angle inside a 2D shape.

Exterior angle: the angle between a side of a 2D shape and a line extended from the neighbouring side.

Unit 6 Quadrilaterals

Quadrilateral: a 2D shape with four straight sides.

Perimeter: the total distance around the edge of a shape.

Vertex: a point where two sides of a shape join.

Vertices: the plural of vertex.

Rectangle: a quadrilateral with four right angles and two pairs of parallel sides; opposite sides are equal.

Square: a quadrilateral with four right angles and two pairs of parallel sides; all sides are equal.

Parallelogram: a quadrilateral with two pairs of parallel sides; opposite sides are equal.

Rhombus: a quadrilateral with two pairs of parallel sides; all sides are equal.

Kite: a quadrilateral with two pairs of equal sides.

Unit 7 Angles

Regular shape: a shape with all side lengths equal and all angles equal.

Unit 8 Loci

Locus: the path that an object follows.

Loci: the plural of locus.

Construction: an accurate drawing of a locus or shape, created using only compasses and a ruler.

Circumference: the perimeter of a circle.

Radius: the distance from the centre of a circle to the circumference.

Arc: part of the circumference of a circle.

Equidistant: of equal distance. A point that is equidistant from two points is the same distance from each point.

The locus of points equidistant from a fixed point is a circle.

The locus of points equidistant from points X and Y is the perpendicular bisector of XY.

Perpendicular bisector: a line that cuts another line in half and crosses that line at right angles.

Perpendicular lines: lines that make an angle of 90°. They are at right angles to each other.

Bisect: to cut in half.

① Simplifying and brackets

This unit will help you to simplify algebraic expressions, including expanding brackets and factorising.

AO1 Fluency check

① Work out

a $2 + 5 - 3$ ....

b $7 - 4 + 1$ ....

c $3 - 8 + 2$ ....

d 2×-4 ....

e -5×3 ....

f -2×-6 ....

② Fill in the missing numbers.

a $5 \times$ .... $= 15$

b $3 \times$ .... $= 3$

c $4 \times$ .... $= -4$

d $2 \times$ .... $= -6$

e $3 \times$ .... $= -9$

f $7 \times$ .... $= -7$

③ Simplify

a $2a + 4a$ ....

b $3a - a$ ....

c $6a + 2a - 3a$ ....

④ Number sense

Match the equivalent calculations.

A $7 \times 2 + 7 \times 3$

i $2(3 + 7)$

B $2 \times 3 + 2 \times 7$

ii $3(7 + 2)$

C $3 \times 7 + 3 \times 2$

iii $7(2 + 3)$

Key points

In algebraic expressions, letters stand for unknown numbers.

$2a$ means '2 lots of a' or $2 \times a$

a	a

These **skills boosts** will help you to simplify expressions.

① Simplifying by collecting like terms

② Multiplying terms

③ Expanding brackets

④ Factorising expressions

You might have already done some work on simplifying and brackets. Before starting the first skills boost, rate your confidence with these questions.

① Simplify $4x + 3y - x + y$

② Simplify $3x \times 4y$

③ Expand $5(x - 6)$

④ Factorise $6x - 15$

How confident are you?

1 Simplifying by collecting like terms

Like terms have exactly the same letter(s) to the same powers.
- $5x$ and $7x$ are like terms.
- $2a$ and $5a^2$ are not like terms.

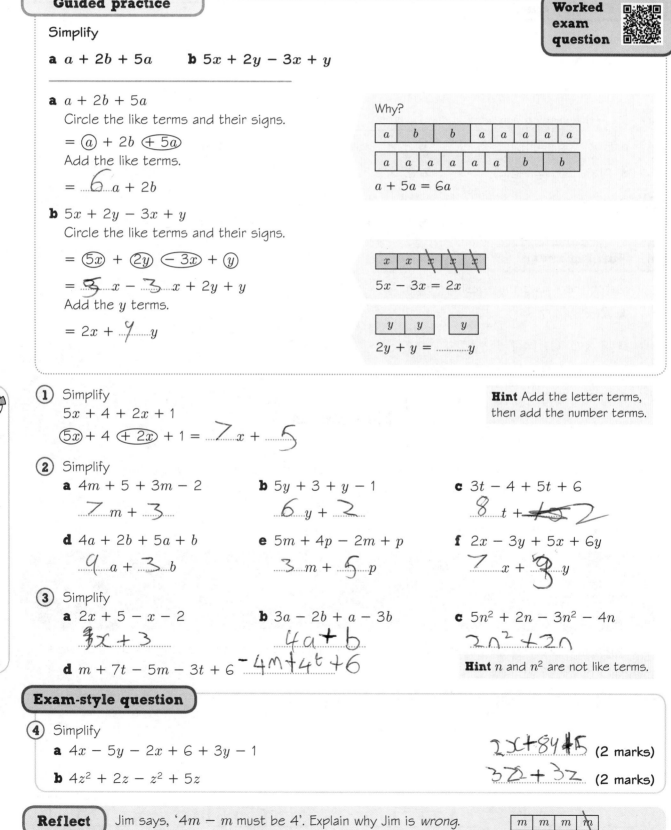

Guided practice

Simplify

a $a + 2b + 5a$ **b** $5x + 2y - 3x + y$

a $a + 2b + 5a$
Circle the like terms and their signs.
= ⓐ + 2b ⊕ 5a
Add the like terms.
= 6 $a + 2b$

Why?

| a | b | | b | a | a | a | a | a |

| a | a | a | a | a | a | | b | b |

$a + 5a = 6a$

b $5x + 2y - 3x + y$
Circle the like terms and their signs.
= ⑤ₓ + ②ᵧ ⊖3x + ⓨ
= 5 $x - 3 x + 2y + y$
Add the y terms.
= $2x + 4 y$

| x | x | $\cancel{x}$ | $\cancel{x}$ | $\cancel{x}$ |

$5x - 3x = 2x$

| y | y | | y |

$2y + y = \underline{\quad} y$

(1) Simplify
$5x + 4 + 2x + 1$

Hint Add the letter terms, then add the number terms.

⑤ₓ + 4 ⊕2x + 1 = 7 $x +$ 5

(2) Simplify
a $4m + 5 + 3m - 2$
7 $m +$ 3

b $5y + 3 + y - 1$
6 $y +$ 2

c $3t - 4 + 5t + 6$
8 $t +$ ~~5~~2

d $4a + 2b + 5a + b$
9 $a +$ 3 b

e $5m + 4p - 2m + p$
3 $m +$ 5 p

f $2x - 3y + 5x + 6y$
7 $x +$ 3 y

(3) Simplify
a $2x + 5 - x - 2$
$\cancel{3}x + 3$

b $3a - 2b + a - 3b$
$4a + b$

c $5n^2 + 2n - 3n^2 - 4n$
$2n^2 + 2n$

d $m + 7t - 5m - 3t + 6$ $\quad -4m + 4t + 6$

Hint n and n^2 are not like terms.

Exam-style question

(4) Simplify
a $4x - 5y - 2x + 6 + 3y - 1$
$2x + 8y - 5$ (2 marks)

b $4z^2 + 2z - z^2 + 5z$
$3z^2 + 3z$ (2 marks)

Reflect Jim says, '$4m - m$ must be 4'. Explain why Jim is *wrong*.

| m | m | m | $\cancel{m}$ |

2 Multiplying terms

To multiply terms with different letters, write them without a × sign in alphabetical order.

$a \times b = ab$ $y \times x = xy$

To multiply terms with the same letter, use a power.

$x \times x = x^2$

To multiply terms with letters and numbers, multiply the numbers, then the letters.

Guided practice

Simplify   **a** $4a \times 2b$ **b** $c \times 2c$

a ④$a \times$ ②b

Multiply the numbers, then the letters.

$= \underline{8} \, ab$

Why?

$4a \times 2b = 4 \times a \times 2 \times b$

$= 4 \times 2 \times a \times b$

$\underbrace{\quad}_{8} \quad \underbrace{\quad}_{ab}$

b $c \times 2c$

Multiply the numbers, then the letters.

$= \underline{2} \, c^{\underline{2}}$

$c = 1 \times c$

$c \times c = c^2$

① Simplify

a ③ × ②a .....6.....a

b $4 \times 3t$ .....$12t$.....   **c** $5 \times 2x$ .....$10x$.....   **d** $3n \times 2$ .....$6n$.....

Hint
3 lots of $2a$

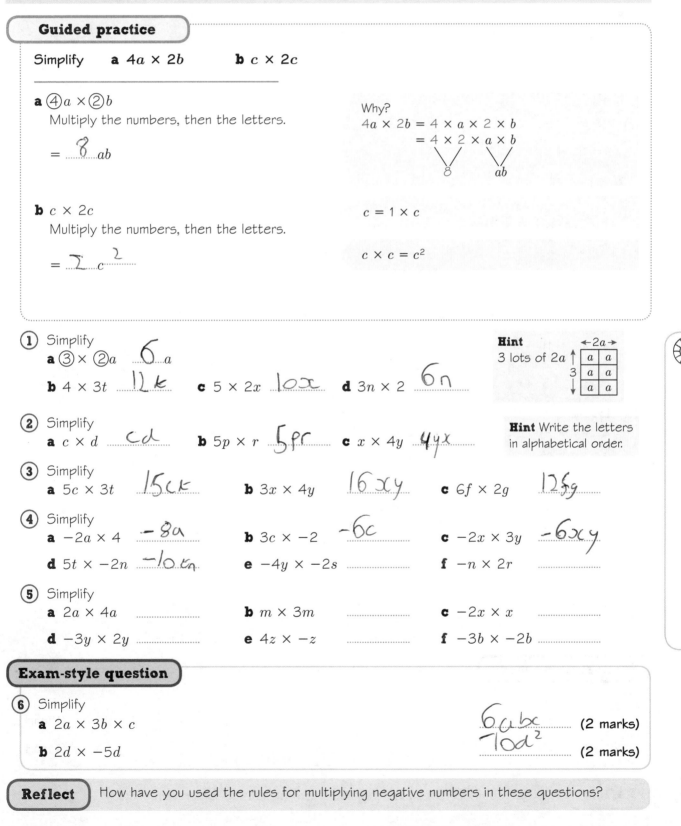

② Simplify

a $c \times d$ .....cd.....   **b** $5p \times r$ .....$5pr$.....   **c** $x \times 4y$ .....$4yx$.....

Hint Write the letters in alphabetical order.

③ Simplify

a $5c \times 3t$ .....$15ck$.....   **b** $3x \times 4y$ .....$16xy$.....   **c** $6f \times 2g$ .....$12fg$.....

④ Simplify

a $-2a \times 4$ .....$-8a$.....   **b** $3c \times -2$ .....$-6c$.....   **c** $-2x \times 3y$ .....$-6xy$.....

d $5t \times -2n$ .....$-10tn$.....   **e** $-4y \times -2s$ ....   **f** $-n \times 2r$ ....

⑤ Simplify

a $2a \times 4a$ ....   **b** $m \times 3m$ ....   **c** $-2x \times x$ ....

d $-3y \times 2y$ ....   **e** $4z \times -z$ ....   **f** $-3b \times -2b$ ....

Exam-style question

⑥ Simplify

a $2a \times 3b \times c$ .....$6abc$..... (2 marks)

b $2d \times -5d$ .....$10d^2$..... (2 marks)

Reflect   How have you used the rules for multiplying negative numbers in these questions?

3 Expanding brackets

To expand brackets, multiply each term inside the bracket by the term outside.

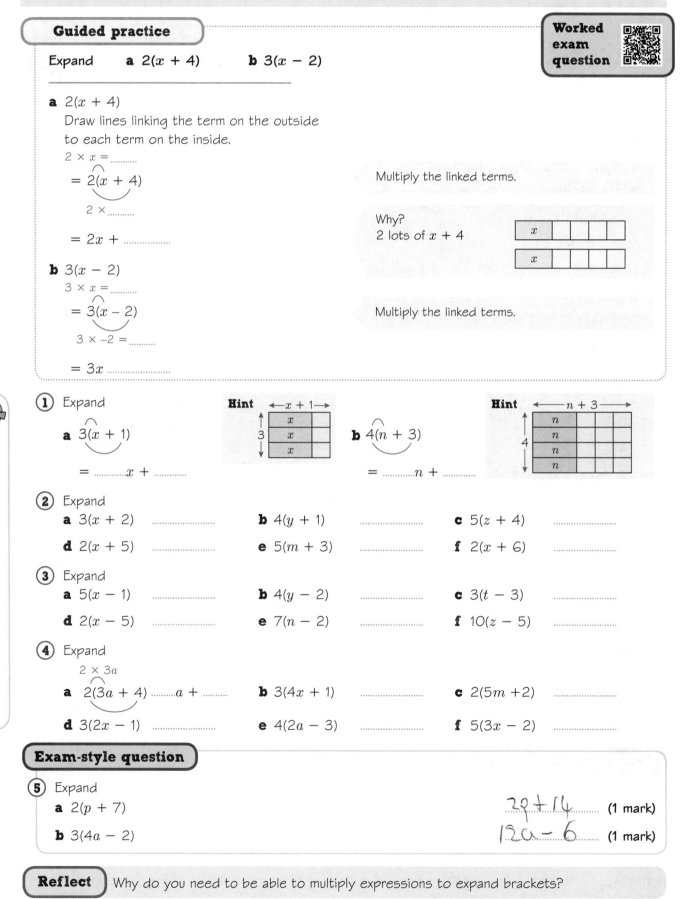

Guided practice

Expand   **a** $2(x + 4)$ **b** $3(x - 2)$

a $2(x + 4)$
Draw lines linking the term on the outside to each term on the inside.

$2 \times x = \text{..........}$

$= 2(x + 4)$

$2 \times \text{..........}$

$= 2x + \text{..............}$

b $3(x - 2)$

$3 \times x = \text{..........}$

$= 3(x - 2)$

$3 \times -2 = \text{..........}$

$= 3x \text{....................}$

Worked exam question

Multiply the linked terms.

Why?
2 lots of $x + 4$

Multiply the linked terms.

① Expand

a $3(x + 1)$

$= \text{..........} x + \text{..........}$

Hint ← $x + 1$ →

b $4(n + 3)$

$= \text{..........} n + \text{..........}$

Hint ← $n + 3$ →

② Expand
a $3(x + 2)$ ....   **b** $4(y + 1)$ ....   **c** $5(z + 4)$ ....
d $2(x + 5)$ ....   **e** $5(m + 3)$ ....   **f** $2(x + 6)$ ....

③ Expand
a $5(x - 1)$ ....   **b** $4(y - 2)$ ....   **c** $3(t - 3)$ ....
d $2(x - 5)$ ....   **e** $7(n - 2)$ ....   **f** $10(z - 5)$ ....

④ Expand

$2 \times 3a$

a $2(3a + 4)$ .... $a +$ ....   **b** $3(4x + 1)$ ....   **c** $2(5m + 2)$ ....
d $3(2x - 1)$ ....   **e** $4(2a - 3)$ ....   **f** $5(3x - 2)$ ....

Exam-style question

⑤ Expand
a $2(p + 7)$ .... $2p + 14$ .... (1 mark)
b $3(4a - 2)$ .... $12a - 6$ .... (1 mark)

Reflect   Why do you need to be able to multiply expressions to expand brackets?

4 Factorising expressions

Factorising puts brackets into an expression.
To factorise an expression, find the highest common factor (HCF) of the terms.

Guided practice

Worked exam question

Factorise

$3x + 6$

Find the highest common factor (HCF) of the two terms. Write it outside the bracket.

$3(\square + \square)$ HCF of 3 and 6 is 3

Draw lines linking the term on the outside to each term on the inside.

$3 \times \square = 3x$

$3(\square + \square)$

$3 \times \square = 6$

$= 3(x + \text{.................})$

HCF = 3

← x + 2 →

x		
x		
x		

Check your answer by expanding.

Expand →

$3(x + 2) = 3x + 6$

← Factorise

---

① Factorise

Hint

a $5x + 10$

$5 \times \square = 5x$

$= 5(\square + \square)$

$5 \times \square = 10$

$= 5(\text{.......} + \text{.......})$

HCF = 5

← x + ... →

x		
x		
x		
x		
x		

b $2x + 6$

$2 \times \square = 2x$

$= 2(\square + \square)$

$2 \times \square = 6$

$= 2(\text{.......} + \text{.......})$

Hint

HCF = 2

← x + ... →

x		
x		

② Factorise

a $3x + 9$ ....

b $4d + 12$ ....

c $2x + 8$ ....

③ Factorise

a $5x - 15$ ....

b $2x - 4$ ....

c $3f - 3$ ....

④ Factorise

a $2x + 2$ ....

b $3a - 3$ ....

c $6n + 8$ ....

Exam-style question

⑤ Factorise

a $5m + 20$ $5(m + 4)$ .... (1 mark)

b $9x - 15$ $3(3x + 5)$ .... (1 mark)

Reflect How can you use expanding to check your factorising is correct?

Practise the methods

Answer this question to check where to start.

Check up

Which of these is correct?

A	B	C	D
$7a - a = 7$ ◯	$2x \times 3y = 5xy$ ◯	$3(x + 4) = 3x + 4$ ◯	$10x + 25 = 5(2x + 5)$ ◯

If you ticked D, go to Q4.　　　　　　　　　If you ticked B, go to Q2 for more practice.

If you ticked A, go to Q1 for more practice.　　If you ticked C, go to Q3 for more practice.

(1) Simplify

Hint $\boxed{x}\ \boxed{x}\ \boxed{\cancel{x}}$

a $3x - x$ ....

b $2y + 3y - y$ ....　**c** $4a + 2b + a$ ....　**d** $2m + 5t - t + m$ ....

(2) Simplify

a $5a \times b$ ....　　　**b** $2m \times 3p$ ....

Hint Multiply 2×3

c $4x \times 2y$ ....　　　**d** $3y \times 7x$ ....

(3) Expand

a $5(a + 4)$　　　　　　**b** $7(x + 2)$ ....　　**c** $4(y - 2)$ ....

　 $= $ ....$a + 20$

(4) Factorise

a $3x + 30$ ....　　**b** $4x - 20$ ....　　**c** $9x + 12$ ....

d $8y + 20$ ....　　**e** $10z - 15$ ....　　**f** $6x - 8$ ....

Exam-style questions

(5) Simplify

a $4x + 3y + 7x - 5y$ .... (2 marks)

b $3x \times 5x$ .... (1 mark)

(6) Expand

a $6(x + 3)$ .... (1 mark)

b $5(2m - 1)$ .... (1 mark)

(7) a Expand

　$3(4x + 5)$ .... (1 mark)

b Factorise

　$7y - 35$ .... (1 mark)

Problem-solve!

(1) Simplify by collecting like terms.

    a $3x + 5 + 2y - 3 + 7y - x$ ....

    b $4f - 2g + 3 - 2f - 5g - 1$ ....

(2) Show that $7x + 5 - 3y + 4y - 8 + 2x \equiv 9x + y - 3$ **Hint** Simplify the left-hand side.

....

(3) Expand

    a $3(1 + x)$ ....   **b** $5(4 - x)$ ....   **c** $2(11 + 5x)$ ....

(4) **a** Explain why the area of the rectangle is given by
the expression $5(x + 2)$

    ....

    b Show that the area of the rectangle can also be written as $5x + 10$

    ....

Exam-style question

(5) Write an expression for the area of the rectangle. $x - 4$

    3

    .... **(2 marks)**

---

(6) Expand

    a $2(x + y)$ ....   **b** $3(2x + z)$ ....   **c** $5(4m - n)$ ....

    d $4(3t - 5w)$........................   **e** $6(2s - 3x)$ ....   **f** $4(5x - 2y)$ ....

(7) Match the equivalent expressions.

    A $4x + 10$ **B** $4x + 8$ **C** $4x + 7$

    i $5x + 10 - x - 3$ ....   **ii** $2(2x + 5)$ ....   **iii** $4(x + 2)$ ....

Exam-style question

(8) **a** Factorise   **b** Expand

    $4x + 14$ .... **(1 mark)** $x(x + 2)$ .... **(1 mark)**

Now that you have completed this unit, how confident do you feel?

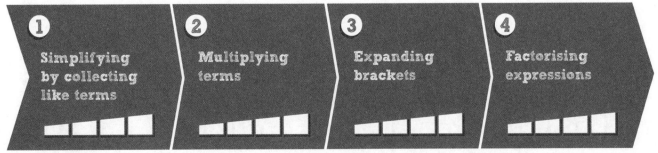

  1 Simplifying by collecting like terms   **2** Multiplying terms   **3** Expanding brackets   **4** Factorising expressions

② Equations, expressions and formulae

This unit will help you to work with equations, expressions and formulae.

AO1 Fluency check

① Work out

a 3^2 .... b 3×2 .... c 7^2 .... d 7×2 ....

② Simplify by collecting like terms.

a $3x - x$ .... b $4x - 2x$ .... c $x^2 + 2x + 3x^2$ ....

③ Expand

a $3(w + 2)$ .... b $5(x - 3)$ .... c $4(2x + 1)$ ....

④ Complete the function machines.

a $3 \rightarrow \boxed{\times 5} \rightarrow 15$

   $3 \leftarrow \boxed{\div \ldots} \leftarrow 15$

b $8 \rightarrow \boxed{\times \ldots} \rightarrow 24$

   $8 \leftarrow \boxed{\div \ldots} \leftarrow 24$

⑤ **Number sense**

a $\frac{1}{4} \times 4 =$ .... b $\frac{3}{4} \times 4 =$ ....

c $\frac{x}{4} \times 4 =$ .... d $\left(\frac{5x - 3}{4}\right) \times 4 =$ ....

Key points

'Solve an equation' means 'work out the value of the letter'.

'Substitute into a formula' means 'replace the letters with the number values you are given'.

These **skills boosts** will help you to solve equations, write expressions and use formulae.

| 1 Solving equations with the letter on one side | 2 Solving equations with the letter on both sides | 3 Writing expressions | 4 Substituting into formulae |

You might have already done some work on equations, expressions and formulae. Before starting the first skills boost, rate your confidence with these questions.

① Solve
$3x - 2 = 13$

② Solve
$5x - 4 = 2x + 5$

③ Write an expression for the cost of n apples at 30p each.

④ $R = IV$
Find R when $I = 3$ and $V = 4$

How confident are you?

1 Solving equations with the letter on one side

In an equation, the expressions on both sides of the equals sign are equal.
To keep both sides equal you need to 'do the same to both sides'.

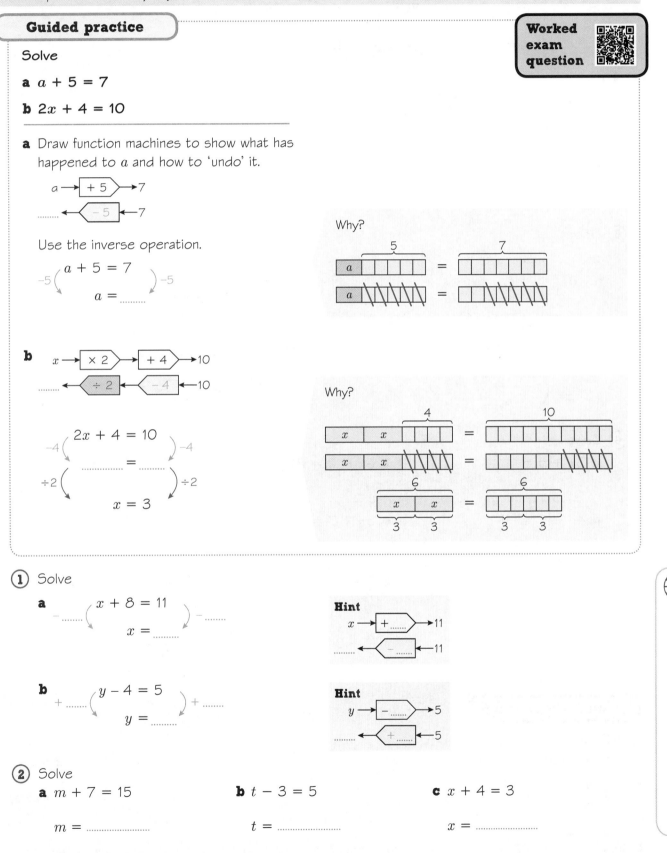

Guided practice

Solve

a $a + 5 = 7$

b $2x + 4 = 10$

a Draw function machines to show what has happened to a and how to 'undo' it.

Use the inverse operation.

$$a + 5 = 7$$
$$a = \text{.........}$$

Why?

b

$$2x + 4 = 10$$
$$\text{.........} = \text{.........}$$
$$x = 3$$

Why?

Worked exam question

① Solve

a
$$x + 8 = 11$$
$$x = \text{.........}$$

Hint

b
$$y - 4 = 5$$
$$y = \text{.........}$$

Hint

② Solve

a $m + 7 = 15$

b $t - 3 = 5$

c $x + 4 = 3$

$m = \text{.....................}$

$t = \text{.....................}$

$x = \text{.....................}$

Unit 2 Equations, expressions and formulae   **9**

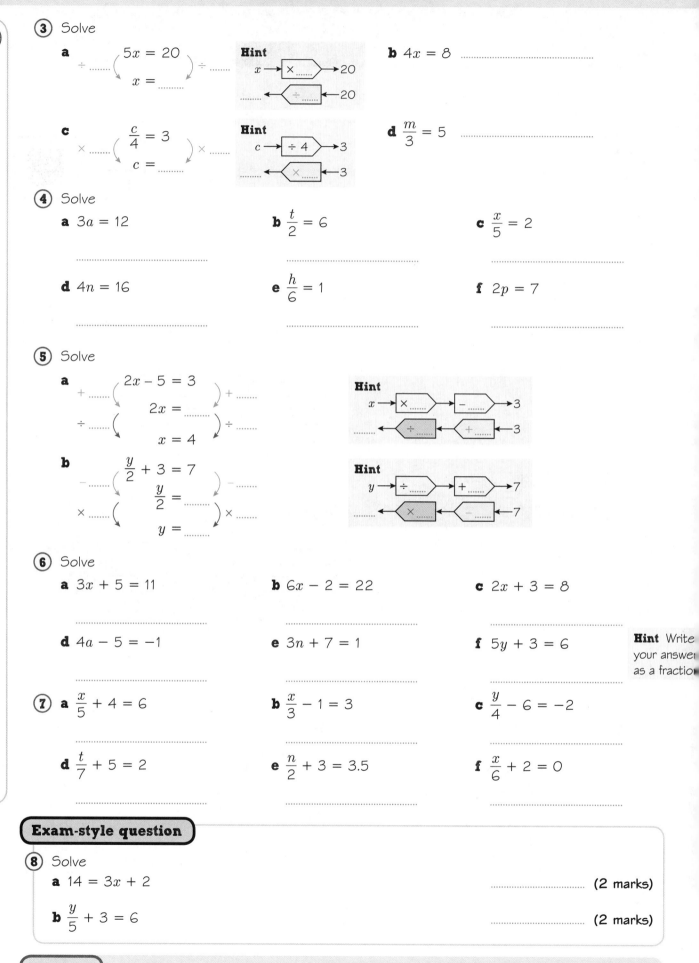

3 Solve

a

$5x = 20$

$\div$ .... $\Bigg($ $\Bigg)$ $\div$ ....

$x =$ ....

Hint

$x \longrightarrow \boxed{\times \text{........}} \longrightarrow 20$

.... $\longleftarrow \boxed{\div \text{........}} \longleftarrow 20$

b $4x = 8$ ....

c

$\times$ .... $\Bigg($ $\dfrac{c}{4} = 3$ $\Bigg)$ $\times$ ....

$c =$ ....

Hint

$c \longrightarrow \boxed{\div 4} \longrightarrow 3$

.... $\longleftarrow \boxed{\times \text{........}} \longleftarrow 3$

d $\dfrac{m}{3} = 5$ ....

4 Solve

a $3a = 12$

b $\dfrac{t}{2} = 6$

c $\dfrac{x}{5} = 2$

d $4n = 16$

e $\dfrac{h}{6} = 1$

f $2p = 7$

5 Solve

a

$+$ .... $\Bigg($ $2x - 5 = 3$ $\Bigg)$ $+$ ....

$\div$ .... $\Bigg($ $2x =$ .... $\Bigg)$ $\div$ ....

$x = 4$

Hint

$x \longrightarrow \boxed{\times \text{........}} \longrightarrow \boxed{- \text{........}} \longrightarrow 3$

.... $\longleftarrow \boxed{\div \text{........}} \longleftarrow \boxed{+ \text{........}} \longleftarrow 3$

b

$-$ .... $\Bigg($ $\dfrac{y}{2} + 3 = 7$ $\Bigg)$ $-$ ....

$\times$ .... $\Bigg($ $\dfrac{y}{2} =$ .... $\Bigg)$ $\times$ ....

$y =$ ....

Hint

$y \longrightarrow \boxed{\div \text{........}} \longrightarrow \boxed{+ \text{........}} \longrightarrow 7$

.... $\longleftarrow \boxed{\times \text{........}} \longleftarrow \boxed{- \text{........}} \longleftarrow 7$

6 Solve

a $3x + 5 = 11$

b $6x - 2 = 22$

c $2x + 3 = 8$

d $4a - 5 = -1$

e $3n + 7 = 1$

f $5y + 3 = 6$

Hint Write your answer as a fraction

7 **a** $\dfrac{x}{5} + 4 = 6$

b $\dfrac{x}{3} - 1 = 3$

c $\dfrac{y}{4} - 6 = -2$

d $\dfrac{t}{7} + 5 = 2$

e $\dfrac{n}{2} + 3 = 3.5$

f $\dfrac{x}{6} + 2 = 0$

Exam-style question

8 Solve

a $14 = 3x + 2$ .... (2 marks)

b $\dfrac{y}{5} + 3 = 6$ .... (2 marks)

Reflect In Q5, Q6, Q7 and Q8, how do function machines help you to use the inverse operations in the correct order?

② Solving equations with the letter on both sides

When an equation has a letter term on both sides of the equals sign, subtract or add to remove the smallest letter term from both sides.

Guided practice

Solve $3x + 5 = 6x + 2$

Worked exam question

Find the smallest letter term: $3x$ or $6x$?

Subtract the smallest letter term from both sides.

$$-3x \left(\begin{array}{c} 3x + 5 = 6x + 2 \\ 5 = \text{......} + 2 \end{array} \right) -3x$$

$$-2 \left(\begin{array}{c} \\ \text{......} = 3x \end{array} \right) -2$$

$$\div \text{......} \left(\begin{array}{c} \\ 1 = x \end{array} \right) \div \text{......}$$

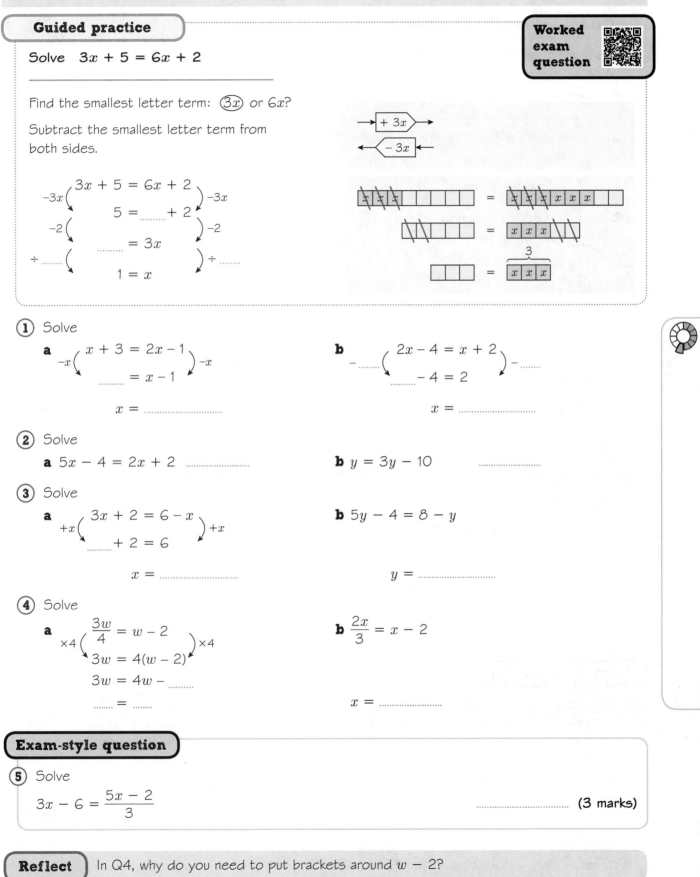

① Solve

a $$-x \left(\begin{array}{c} x + 3 = 2x - 1 \\ \text{......} = x - 1 \end{array} \right) -x$$

$x = $ ....

b $$- \text{......} \left(\begin{array}{c} 2x - 4 = x + 2 \\ \text{......} - 4 = 2 \end{array} \right) - \text{......}$$

$x = $ ....

② Solve

a $5x - 4 = 2x + 2$ ....

b $y = 3y - 10$ ....

③ Solve

a $$+x \left(\begin{array}{c} 3x + 2 = 6 - x \\ \text{......} + 2 = 6 \end{array} \right) +x$$

$x = $ ....

b $5y - 4 = 8 - y$

$y = $ ....

④ Solve

a $$\times 4 \left(\begin{array}{c} \dfrac{3w}{4} = w - 2 \\ 3w = 4(w - 2) \end{array} \right) \times 4$$

$3w = 4w - \text{......}$

$\text{......} = \text{......}$

b $\dfrac{2x}{3} = x - 2$

$x = $ ....

Exam-style question

⑤ Solve

$$3x - 6 = \frac{5x - 2}{3}$$

.... **(3 marks)**

Reflect In Q4, why do you need to put brackets around $w - 2$?

3 Writing expressions

To write an expression, use letters to stand for values you don't know.
The letters are called variables because their values can change or vary.

Guided practice

There are x tennis balls in a box.
Write an expression for the number of tennis balls in

a 3 boxes   **b** 2 boxes plus 3 extra balls.

___

a How do you get from 1 box to 3 boxes?

$\times 3 \left(\begin{array}{cc} 1\text{ box} & x\text{ balls} \\ \text{......... boxes} & 3x\text{ balls} \end{array} \right) \times 3$

Why?
x	x	x

b Go from 1 box to 2 boxes.

$\times 2 \left(\begin{array}{cc} 1\text{ box} & x\text{ balls} \\ 2\text{ boxes} & \text{......... balls} \end{array} \right) \times 2$

Add 3 extra.

$2x +$ ....

Why?
x	x		

1 Write expressions for

a 3 more than n ....

Hint
n			

b 4 more than y ....

c 5 lots of m ....

Hint
m	m	m	m	m

d 3 lots of x ....

e 3 less than z ....

f half of t ....

2 There are n biscuits in a pack. Write an expression for the number of biscuits in

a 2 packs ....

b 5 packs plus 6 biscuits ....

c 3 packs with 4 biscuits taken out ....

3 Manesh earns £5.20 per hour.

Write an expression for the amount he earns in n hours.

$\times n \left(\begin{array}{cc} 1\text{ hour} & £5.20 \\ n\text{ hours} & \text{...............} \end{array} \right) \times \text{........}$

Exam-style question

4 **a** Alix buys some festival tickets and two car park tickets.
Write an expression for the total cost.

| Festival tickets | £48 |
| Car park tickets | £20 |

.... (1 mark)

b Alix pays £376 in total. How many festival tickets does she buy?

.... (1 mark)

Reflect

How can drawing diagrams help you write expressions?

4 Substituting into formulae

To use a formula, substitute (swap) the letters for the values you are given.

Guided practice

Worked exam question

$d = st$, where d is distance, s is speed and t is time.

a Find d when $s = 5$ and $t = 8$

b Find s when $d = 30$ and $t = 2$

---

a Swap s and t for the values given.

$d = st$

$d = 5 \times$ ....

$=$ ....

> st means $s \times t$
> $s = 5$ and $t = 8$

b Swap d and t for the values given.

$d = st$

$30 = s \times$ ....

$30 =$ .... s

Solve the equation.

$s =$ ....

> $d = 30$ and $t = 2$

(1) Find the value of each expression when $n = 4$

a $n - 2$ ....   **b** $n + 5$ ....   **c** $2n$ ....   **d** $\dfrac{n}{2}$ ....

e $2n + 1$ ....   **f** $13 - 3n$ ....   **g** $\dfrac{n}{2} + 3$ ....   **h** $\dfrac{3n}{4}$ ....

(2) $F = ma$, where F is force, m is mass and a is acceleration.

a Find F when $m = 5$ and $a = 3$

b Find F when $m = 10$ and $a = 0.3$

....

....

(3) $P = \dfrac{F}{A}$

a Find P when $F = 18$ and $A = 2$

b Find P when $F = 14$ and $A = 4$

....

....

(4) $y = ax^2$

a Find y when $a = 3$ and $x = 2$

b Find y when $a = \frac{1}{2}$ and $x = 4$

....

....

Exam-style question

(5) a $s = \dfrac{d}{t}$

Find d when $s = 50$ and $t = 3$

.... (2 marks)

b $v = u + at$

Find u when $v = 22$, $a = 3$ and $t = 4$

.... (2 marks)

Reflect When have you used your equation-solving skills in these questions?

Practise the methods

Answer this question to check where to start.

Check up

$V = mx^2$ Find V when $m = 3$ and $x = 4$. Tick the correct answer.

Ⓐ ◯
$V = 3 \times 16$
$= 48$

Ⓑ ◯
$V = 3 \times 8$
$= 24$

Ⓒ ◯
$V = 144$

Ⓓ ◯
$V = 316$

If you ticked A go to Q2.

If you ticked B, C or D go to Q1 for more practice.

① Work out

  a $4m$ when $m = 5$ ....

  b kn when $k = 2$ and $n = 6$ ....

  c $5x^2$ when $x = 3$ ....

  d rt^2 when $r = 7$ and $t = 2$ ....

② A large lorry has n wheels.

Write expressions for the number of wheels on

  a 2 of these lorries ....

  b 5 lorries plus 3 spare wheels ....

③ $y = 2x - 3$

  a Find the value of y when $x = 5$

  b Find the value of x when $y = 15$ ....

  $y =$ ....

  $x =$ ....

Exam-style questions

④ Lucy earns £80 per day, plus £5 for each dress she sells.

  a Write an expression for the amount she earns on a day when she sells x dresses.

  .... (1 mark)

  b One day she earned £115. How many dresses did she sell?   .... (2 marks)

⑤ Solve

  a $9 + 3a = 15$ .... (1 mark)   **b** $\frac{4x}{5} = 8$ .... (1 mark)

⑥ Solve

  a $2m + 6 = 5m - 9$

  b $\frac{2x + 1}{3} = x - 2$

  $m =$ ....

  $x =$ ....

Exam-style question

⑦ $s = ut + \frac{1}{2}at^2$

Find s when $u = 5$, $t = 6$ and $a = 2$ .... (3 marks)

Problem-solve!

(1) Write a formula for the cost C of x cinema tickets at £6.50 each.

.... (2 marks)

(2) In the formula $v^2 - u^2 = 2as$
v is final velocity, u is initial velocity, a is acceleration and s is distance travelled.
Find a when $v = 12$, $u = 4$ and $s = 10$

.... (3 marks)

(3) Solve

a $2 + 5x = 17$ **b** $8 - 3y = 2$ **c** $4x + 7 = 3$

$x = $ .... $y = $ .... $x = $ ....

d $\dfrac{x}{4} = -2$ ....   **e** $6x = 3$ ....   **f** $10x + 5 = 7$ ....

$x = $ .... $x = $ .... $x = $ ....

(4) The rectangle has perimeter 34 cm.

Find x.

.... (2 marks)

(5) The rectangle has area 30 cm².

Find the length of the longer side.

.... (2 marks)

(6) Find x.

.... (2 marks)

(7) An electrician uses this formula to calculate his charges:

$C = 30 + 20h$

a What does C stand for?   ....

b What does h stand for?   ....

c How much would he charge for a job that takes 5 hours?   ....

Now that you have completed this unit, how confident do you feel?

1 Solving equations with the letter on one side

2 Solving equations with the letter on both sides

3 Writing expressions

4 Substituting into formulae

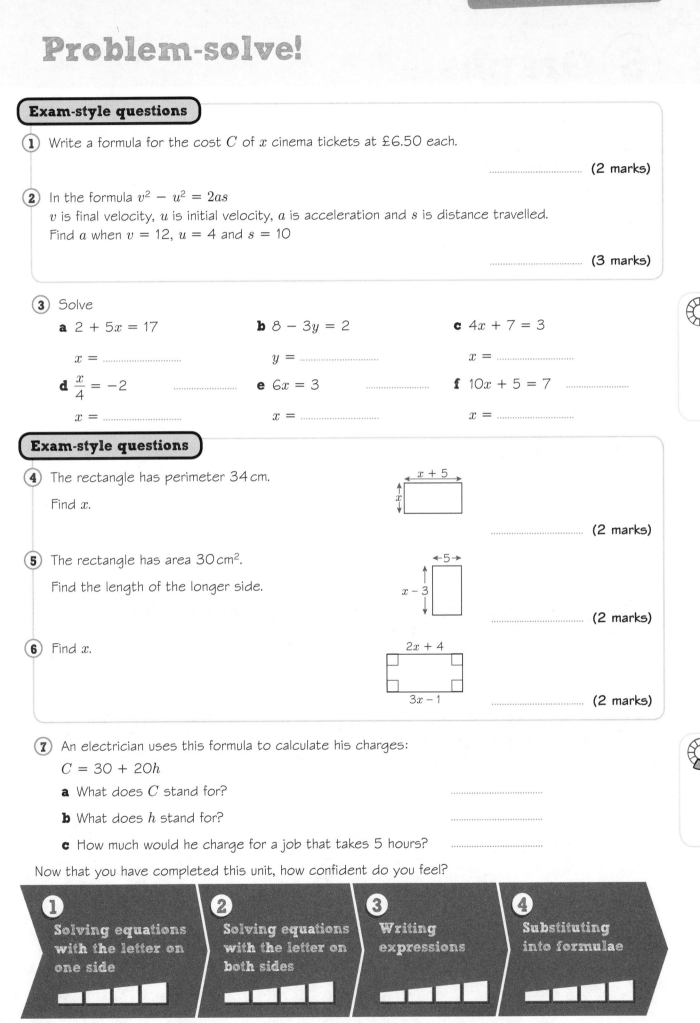

③ Graphs

This unit will help you to draw and interpret different types of graph.

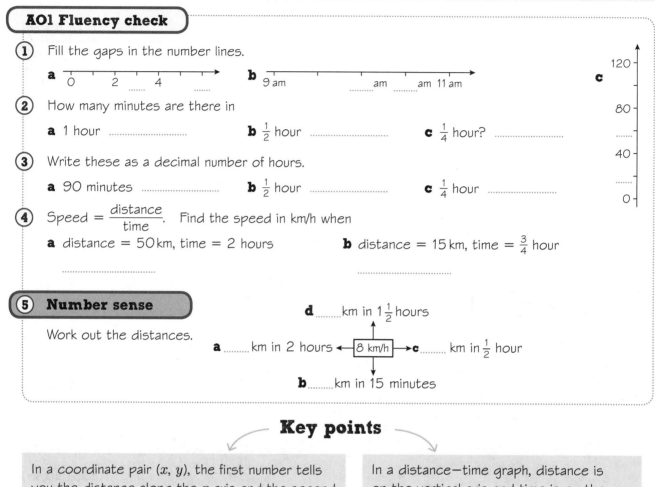

AO1 Fluency check

① Fill the gaps in the number lines.

a 0   2 ...... 4 ......

b 9 am .... am .... am 11 am

c 120 — 80 — 40 — 0 —

② How many minutes are there in

a 1 hour ....

b $\frac{1}{2}$ hour ....

c $\frac{1}{4}$ hour? ....

③ Write these as a decimal number of hours.

a 90 minutes ....

b $\frac{1}{2}$ hour ....

c $\frac{1}{4}$ hour ....

④ Speed = $\frac{\text{distance}}{\text{time}}$. Find the speed in km/h when

a distance = 50 km, time = 2 hours

b distance = 15 km, time = $\frac{3}{4}$ hour

⑤ **Number sense**

Work out the distances.

d .... km in 1$\frac{1}{2}$ hours

a .... km in 2 hours ← 8 km/h → c .... km in $\frac{1}{2}$ hour

b .... km in 15 minutes

Key points

In a coordinate pair (x, y), the first number tells you the distance along the x-axis and the second number tells you the distance along the y-axis.

In a distance–time graph, distance is on the vertical axis and time is on the horizontal axis.

These **skills boosts** will help you to draw and interpret different types of graph.

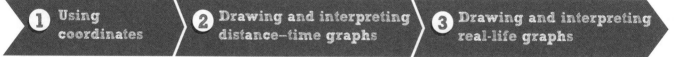

① Using coordinates ② Drawing and interpreting distance–time graphs ③ Drawing and interpreting real-life graphs

You might have already done some work on graphs. Before starting the first skills boost, rate your confidence with these questions.

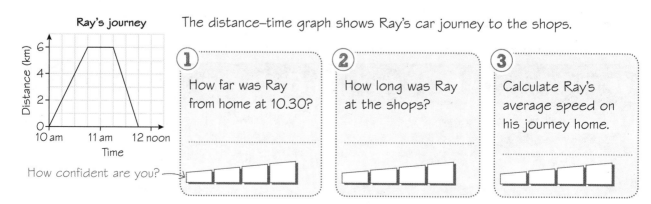

Ray's journey

The distance–time graph shows Ray's car journey to the shops.

① How far was Ray from home at 10.30?

② How long was Ray at the shops?

③ Calculate Ray's average speed on his journey home.

How confident are you?

1 Using coordinates

On a coordinate grid, the x-axis is horizontal and the y-axis is vertical.

Guided practice

Write down the coordinates of A, B, C and D.

From point A
• look down to the x-axis for the x-value
• look across to the y-axis for the y-value.

A = (3, 2)

        x y

B = (−4, ...)

C = (−.........., −..........)

D = (.........., ...)

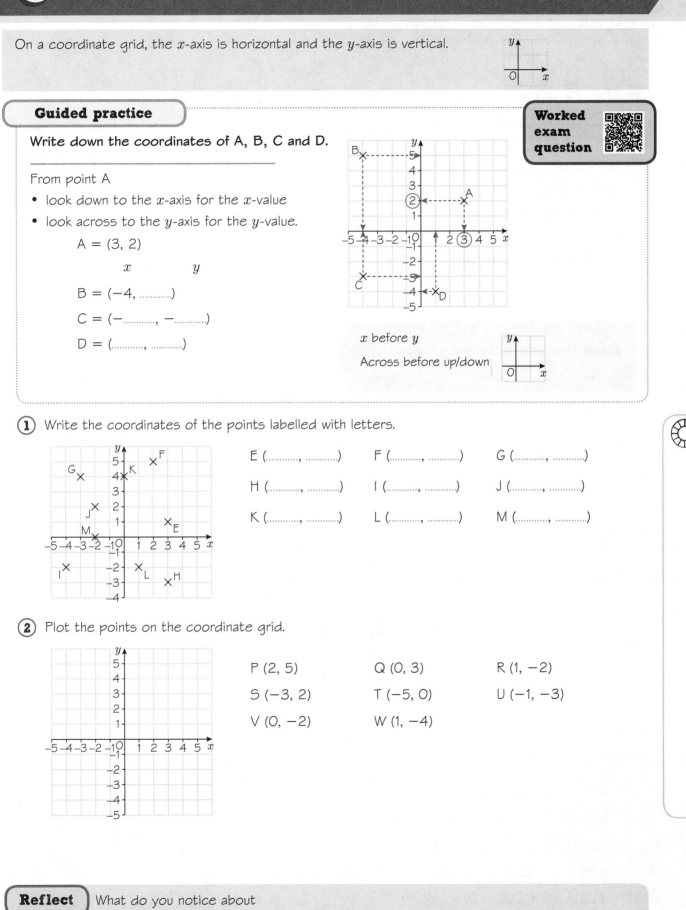

x before y

Across before up/down

① Write the coordinates of the points labelled with letters.

E (.........., ...)   F (.........., ...)   G (.........., ...)

H (.........., ...)   I (.........., ...)   J (.........., ...)

K (.........., ...)   L (.........., ...)   M (.........., ...)

② Plot the points on the coordinate grid.

P (2, 5)   Q (0, 3)   R (1, −2)

S (−3, 2)   T (−5, 0)   U (−1, −3)

V (0, −2)   W (1, −4)

Reflect What do you notice about
• the y-coordinates of points on the x-axis
• the x-coordinates of points on the y-axis?

2 Drawing and interpreting distance–time graphs

In a distance–time graph, a horizontal line shows no change in distance, so the person or object is not moving.

- Speed = $\dfrac{\text{distance}}{\text{time}}$
- Average speed = $\dfrac{\text{total distance}}{\text{total time}}$

Guided practice

The graph shows Mia's bike ride.

a What time did Mia stop to rest?

b How long did she rest for?

c Calculate her speed in the first hour.

d Calculate her speed after the rest.

Mia's bike ride

Horizontal line: not moving

Worked exam question

_____

a Read the time at the start of the horizontal line.

3 pm

b Read the time at the end of the horizontal line.

3 pm to .... pm = ... minutes

c How far did she travel in the first hour?

.... km in 1 hour = ... km/h

Speed = number of kilometres in 1 hour

d Speed = $\dfrac{\text{distance}}{\text{time}}$

= $\dfrac{\text{................}}{\text{................}}$

= .... km/h

Distance from 12.5 to 17.5 = ☐ km
Time from 3.30 pm to 5.30 pm = ☐ hours

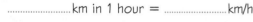

1 The graph shows Dan's walk.

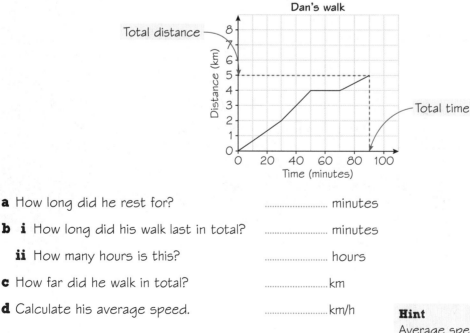

Dan's walk

Total distance

Total time

a How long did he rest for? .... minutes

b i How long did his walk last in total? .... minutes

ii How many hours is this? .... hours

c How far did he walk in total? .... km

d Calculate his average speed. .... km/h

Hint
Average speed = $\dfrac{\text{total distance}}{\text{total time}}$
Use time in hours.

(2) Draw a distance—time graph to show Nusrat's journey to the shops and back home.

Nusrat leaves home at 10 am. Plot this point.

She walks to the shop 2 km away. She arrives at the shop at 10.30 am. Plot this point and draw the line.

She spends 15 minutes at the shop. Draw the horizontal line.

She walks home and arrives at 11.15 am. Draw the line to this point.

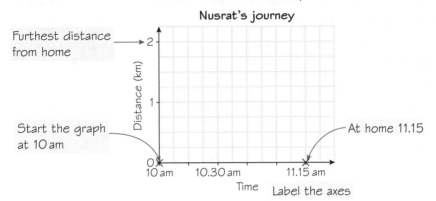

Nusrat's journey

Furthest distance from home

Distance (km)

Start the graph at 10 am

At home 11.15

Time   Label the axes

(3) The graph shows Sean's visit to his friend's house.

a What is the distance to his friend's house?

.... km

b How long did Sean take to reach his friend's house?

Give your answer as a decimal.

.... hours

c Calculate his speed on the way to his friend's house.

.... km/h

d How long did he stay at his friend's house?

.... hours

e What time did he get home? .... pm

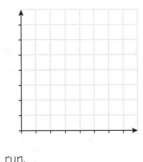

Sean's visit

Distance (km)

Time

Exam-style question

(4) Amy leaves home at 9 am.

She runs 3 km in $\frac{1}{2}$ hour. Then she stops to rest for 10 minutes before running home in 20 minutes.

a Draw a distance—time graph for her run.

(3 marks)

b Calculate her average speed for the whole run. .... (2 marks)

Reflect What are the units of speed when distance is in kilometres and time is in hours?

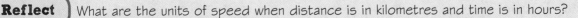

3 Drawing and interpreting real-life graphs

In a graph showing how a quantity changes over time, time is on the horizontal axis.

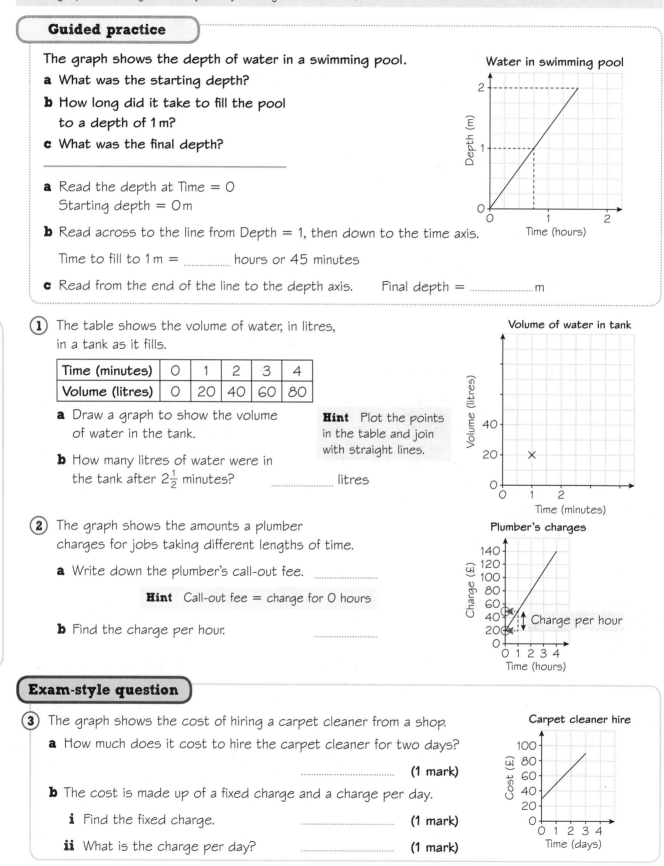

Guided practice

The graph shows the depth of water in a swimming pool.

a What was the starting depth?

b How long did it take to fill the pool to a depth of 1 m?

c What was the final depth?

_____

a Read the depth at Time = 0
Starting depth = 0 m

b Read across to the line from Depth = 1, then down to the time axis.

Time to fill to 1 m = ... hours or 45 minutes

c Read from the end of the line to the depth axis.   Final depth = ... m

① The table shows the volume of water, in litres, in a tank as it fills.

Time (minutes)	0	1	2	3	4
Volume (litres)	0	20	40	60	80

a Draw a graph to show the volume of water in the tank.

Hint Plot the points in the table and join with straight lines.

b How many litres of water were in the tank after $2\frac{1}{2}$ minutes?   .... litres

② The graph shows the amounts a plumber charges for jobs taking different lengths of time.

a Write down the plumber's call-out fee.   ....

Hint Call-out fee = charge for 0 hours

b Find the charge per hour.   ....

Exam-style question

③ The graph shows the cost of hiring a carpet cleaner from a shop.

a How much does it cost to hire the carpet cleaner for two days?

.... **(1 mark)**

b The cost is made up of a fixed charge and a charge per day.

i Find the fixed charge.   .... **(1 mark)**

ii What is the charge per day?   .... **(1 mark)**

Reflect   How did you use coordinates to plot the graph in Q1?

Practise the methods

Answer this question to check where to start.

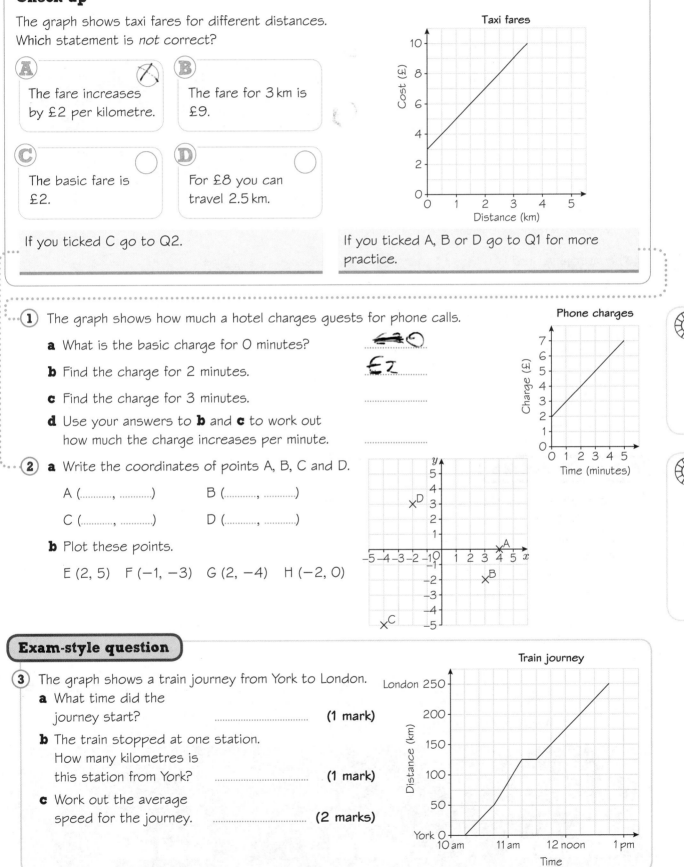

Check up

The graph shows taxi fares for different distances.
Which statement is *not* correct?

A ⊘
The fare increases by £2 per kilometre.

B
The fare for 3 km is £9.

C ◯
The basic fare is £2.

D ◯
For £8 you can travel 2.5 km.

Taxi fares

If you ticked C go to Q2.

If you ticked A, B or D go to Q1 for more practice.

① The graph shows how much a hotel charges guests for phone calls.

a What is the basic charge for 0 minutes? ~~£9~~

b Find the charge for 2 minutes. £2

c Find the charge for 3 minutes. ....

d Use your answers to **b** and **c** to work out how much the charge increases per minute. ....

Phone charges

② **a** Write the coordinates of points A, B, C and D.

A (..........., ...)   B (..........., ...)

C (..........., ...)   D (..........., ...)

b Plot these points.

E (2, 5)   F (−1, −3)   G (2, −4)   H (−2, 0)

Exam-style question

③ The graph shows a train journey from York to London.

a What time did the journey start? .... **(1 mark)**

b The train stopped at one station. How many kilometres is this station from York? .... **(1 mark)**

c Work out the average speed for the journey. .... **(2 marks)**

Train journey

Problem-solve!

① The graph shows the depth of water in a tank.

   a What was the initial depth of water in the tank?

     90 .... (1 mark)

   b By drawing a line on the graph, predict when the tank will be empty.

     50 .... (2 marks)

Depth of water

② Plot the points (1, 2), (−2, 2) and (−2, −1) on a coordinate grid.
These are three vertices of a square.
Write down the coordinates of the fourth vertex.   (.........., ...)

③ Water is poured into these two cylinders at the same steady rate.

   a Which cylinder fills more quickly?   ....

   b The graph shows how the depth of water in the cylinders changes over time.
      Match each cylinder to its line on the graph.

      Cylinder A → line ....

      Cylinder B → line ....

Cylinder A   Cylinder B

④ The graph shows the cost of hiring a camper van from Campers R Us.
Ready Camp hires camper vans for £150 per day with no basic charge.
Bob wants to hire a camper van for a fortnight.
Which company should he choose?

Show workings to explain your answer.

Camper van hire

.... (3 marks)

Now that you have completed this unit, how confident do you feel?

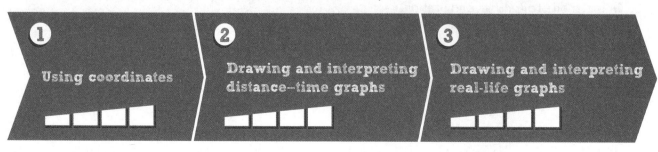

1 Using coordinates

2 Drawing and interpreting distance–time graphs

3 Drawing and interpreting real-life graphs

④ Sequences

This unit will help you to generate sequences and find missing terms using term-to-term rules.

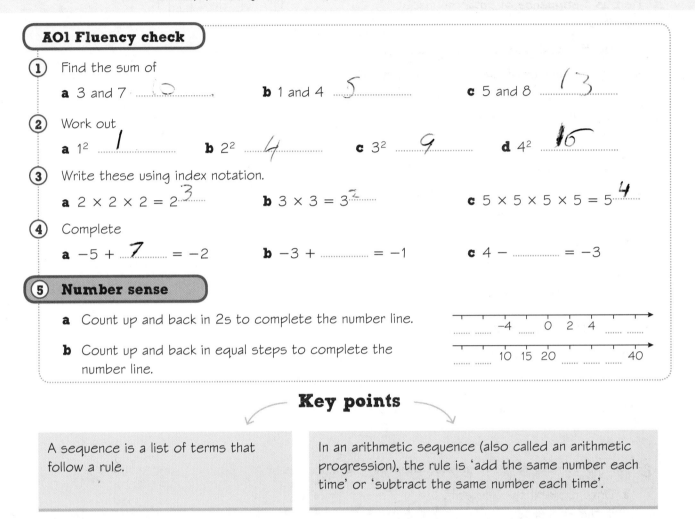

A01 Fluency check

① Find the sum of

a 3 and 7 ....10....   **b** 1 and 4 ....5....   **c** 5 and 8 ....13....

② Work out

a 1^2 ....1....   **b** 2^2 ....4....   **c** 3^2 ....9....   **d** 4^2 ....16....

③ Write these using index notation.

a $2 \times 2 \times 2 = 2^3$ **b** $3 \times 3 = 3^2$ **c** $5 \times 5 \times 5 \times 5 = 5^4$

④ Complete

a $-5 + 7 = -2$ **b** $-3 + \text{.....} = -1$ **c** $4 - \text{.....} = -3$

⑤ **Number sense**

a Count up and back in 2s to complete the number line.

b Count up and back in equal steps to complete the number line.

(number line: −4 0 2 4)

(number line: 10 15 20 ... 40)

Key points

A sequence is a list of terms that follow a rule.

In an arithmetic sequence (also called an arithmetic progression), the rule is 'add the same number each time' or 'subtract the same number each time'.

These **skills boosts** will help you to generate sequences and find missing terms.

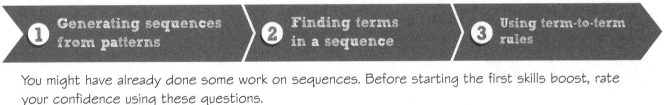

1 Generating sequences from patterns   **2** Finding terms in a sequence   **3** Using term-to-term rules

You might have already done some work on sequences. Before starting the first skills boost, rate your confidence using these questions.

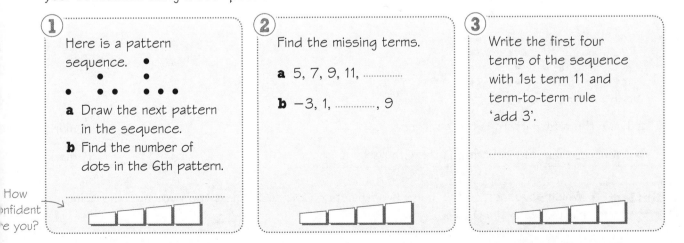

①

Here is a pattern sequence.

•
• •   •
• • •   • • •

a Draw the next pattern in the sequence.

b Find the number of dots in the 6th pattern.

②

Find the missing terms.

a 5, 7, 9, 11, ...

b −3, 1, ..., 9

③

Write the first four terms of the sequence with 1st term 11 and term-to-term rule 'add 3'.

How confident are you?

1 Generating sequences from patterns

Guided practice

Here is a sequence of patterns made from tiles.

Pattern number   1   2   3

a Draw the next pattern in the sequence.

b Find the number of tiles in the 7th pattern.

---

a See how the pattern grows.

Pattern number 4

Every pattern is 2 tiles high.
The pattern gets 1 tile wider each time.

b Write down the number of tiles in each pattern.
Continue the sequence.

Pattern number   1   2   3   4   5   6   7

               2   4   6   8   14

The sequence is the multiples of 2.

The 7th pattern has .... tiles.

① This pattern sequence generates the square numbers.

   a Draw the next two patterns in the sequence.

   b Write the first five numbers in the sequence.   ...., ..., ..., ..., ...

   c Show that the sequence is the same as $1^2, 2^2, 3^2, 4^2, 5^2, \ldots$

② This pattern sequence generates the triangular numbers.

   a Draw the next two patterns in the sequence.

   b Write the first six numbers in the sequence.
     Work out what you add each time to get the next term in the sequence.

     1,   3, ..., ..., ..., ...
     +2 + ... + ... + ... + ...

Exam-style question

③ Here is a sequence of patterns made from counters.

Pattern number   1   2   3

   a Draw the next pattern in the sequence.   **(1 mark)**

   b Find the number of counters in the 6th pattern.   ....   **(1 mark)**

Reflect   Which sequence on this page is arithmetic? (To get the next term you add the same number each time.)

2 Finding terms in a sequence

To find a missing term, work out the rule to get the next term. This is called the term-to-term rule.

Guided practice

Worked exam question

Find the next term in each sequence.

a 1, 5, 9, 13, ...

b 8, 5, 2, −1, ...

---

a Work out what you add or subtract each time. Follow the pattern.

1, 5, 9, 13, ...
+4 + ... + ... + ...

The terms are getting larger (increasing).

b 8, 5, 2, −1, ...
−3 − ..... − ..... − .....

The terms are getting smaller (decreasing).

① Find the next three terms in each sequence.

a 3, 9, 15, 21, ..., ..., ...
+..... +..... +..... +..... +..... +.....

b −5, −3, −1, 1, ..., ..., ...
+..... +..... +..... +..... +..... +.....

c 59, 61, 63, 65, ..., ..., ...

d 14, 21, 28, 35, ..., ..., ...

② Find the next three terms in each sequence.

a 74, 64, 54, 44, ..., ..., ...
−..... −..... −..... −..... −.....

b −2, −5, −8, −11, ..., ..., ...
−..... −..... −..... −..... −.....

c −15, −12, −9, −6, ..., ..., ...

d 8, 4, 0, −4, ..., ..., ...

③ Write the rule for each arithmetic sequence. Find the missing terms.

a 2, 5, ..., 11, ...   Rule: add ....

b ...., 7, 11, ..., ...   Rule: add ....

c −5, 0, ..., ..., 15   Rule: add ....

d 17, 8, ..., −10, ...   Rule: subtract ....

④ Describe each sequence and find the 10th term.

a 4, 8, 12, 16, ...   Multiples of ....   10th term = ...

b 10, 20, 30, 40, ...   Multiples of ....   10th term = ...

c −5, −10, −15, −20, ...   Multiples of −..................   10th term = ...

Exam-style question

⑤ Cherry saves £7 each week.

a Complete the table of her savings.

Week	1	2	3	4	5
Savings (£)	7	14	......	......	......

(1 mark)

b How much money does she save in eight weeks? .... (1 mark)

c How many weeks does it take her to save £84? .... (1 mark)

Reflect   Look at the sequences in Q4. Are all sequences of multiples arithmetic sequences?

3 Using term-to-term rules

You can generate a sequence from the 1st term and the term-to-term rule.

Guided practice

A sequence has 1st term 16 and term-to-term rule 'subtract 5'.

a Write the first four terms.

b Find the 7th term.

a Write the 1st term and the rule.

16, ..., ..., ...
  −5   −5   −5

b Continue the sequence to the 7th term.

16, 11, 6, 1, −4, −..........., ... ← 7th term

---

① Write the first four terms of each sequence.

a 1st term 3, term-to-term rule 'add 7'   ...., ..., ..., ...

b 1st term 18, term-to-term rule 'subtract 2   ...., ..., ..., ...

c 1st term 20, term-to-term rule 'subtract 8'   ...., ..., ..., ...

d 1st term −3, term-to-term rule 'add 4'   ...., ..., ..., ...

e 1st term −1, term-to-term rule 'subtract 2'   ...., ..., ..., ...

f 1st term 20, term-to-term rule 'add 25'   ...., ..., ..., ...

② Write the first four terms of each sequence.

a 2nd term 5, term-to-term rule 'add 2'

..........., 5, ..., ...

b 3rd term 10, term-to-term rule 'subtract 3'

..........., ..., 10, ...

c 2nd term 11, term-to-term rule 'subtract 2'

d 4th term 4, term-to-term rule 'add 3'

③ **a** A sequence has 1st term 6 and term-to-term rule 'add 3'. Find the 9th term.

b A sequence has 1st term 24 and term-to-term rule 'subtract 4'. Find the 8th term.

Exam-style question

④ The rule for a Fibonacci sequence is:

'The next term in the sequence is the sum of the previous two terms.'

Write the next two terms in this Fibonacci sequence.

1, 1, 2, 3, 5, ..., ...   (2 marks)

Reflect

In Q2, how did you use inverse operations to find the missing terms?

Practise the methods

Answer this question to check where to start.

Check up

Find the number of tiles in the 8th pattern in this sequence.

Tick the correct answer.

A ◯ 26

B ◯ 29

C ◯ 33

D ◯ 25

If you ticked B go to Q2.

If you ticked A, C or D go to Q1.

1. Here is a sequence of patterns made from sticks.

   a Draw the next two patterns.

   b Write down the sequence of numbers of sticks. ....

   c Continue the sequence to find the number of sticks in the 6th pattern. ....

   d Sasha says, 'There are 7 sticks in the 3rd pattern, so there will be 7 × 2 = 14 sticks in the 6th pattern'. Is she correct?

   ....

2. Find the next term in each sequence. **a** −6, −2, 2, ... **b** 15, 7, −1, ...

Exam-style question

3. Find the missing terms in these arithmetic sequences.

   a 18, ..., 34, 42 (1 mark) **b** ...., 34, 23, 12 (1 mark)

4. **a** Complete this sequence of calculations.

   1 × 1   2 × 2   3 × 3   .... × .... × .... × ....

   1   4   ....

   b What is the special name for this number sequence? ....

5. Work out the 8th term in the sequence of multiples of 6. ....

Exam-style questions

6. Write the first five terms of the sequence with 1st term 10 and term-to-term rule 'subtract 3'.

   ....

   (2 marks)

7. The table shows taxi fares for different distances.

Distance (km)	2	5	8	11
Fare (£)	3	9	15	21

   Work out the fare for a 17 km trip. .... (2 marks)

Problem-solve!

1 Find the missing terms in these sequences.

a 1, 2.5, ..., 5.5   **(1 mark)**   **b** 1, ..., $\frac{1}{2}$, $\frac{1}{4}$ **(1 mark)**

2 For each arithmetic sequence, write the 1st term and the term-to-term rule.

     a 4, 7, 10, 13, …   ....   **b** 12, 6, 0, −6, …   ....

     c −8, −5, −2, 1, …   ....   **d** −2, −7, −12, −17, …   ....

3 Here are the first four terms in an arithmetic sequence.

     37, 30, 23, 16, …

Find the first term that is less than zero.   .... **(1 mark)**

4 An arithmetic sequence begins

     7, 9, 11, 13, 15, …

Is 100 a term in this sequence? Explain how you know.

.... **(2 marks)**

5 The rule for a Fibonacci sequence is:

     'The next term in the sequence is the sum of the previous two terms.'

Write the next three terms in this Fibonacci sequence.

0, 1, ..., ..., ...   **(3 marks)**

6 Match each description to a sequence.

     A Cube numbers   **i**   1, 4, 9, 16, 25, …

     B Fibonacci   **ii**   1, 8, 27, 64, …

     C Square numbers   **iii**   3, 5, 8, 13, 21, …

     D Powers of 2   **iv**   1, 3, 6, 10, 15, …

     E Triangular numbers   **v**   2, 4, 8, 16, …

7 A shop stacks tins like this.

     a How many tins are there in a stack with seven rows?   .... **(1 mark)**

     b How many rows do 21 tins make?   .... **(1 mark)**

Now that you have completed this unit, how confident do you feel?

1 Generating sequences from patterns

2 Finding terms in a sequence

3 Using term-to-term rules

⑤ Triangles

This unit will help you to find angles and lengths in triangles.

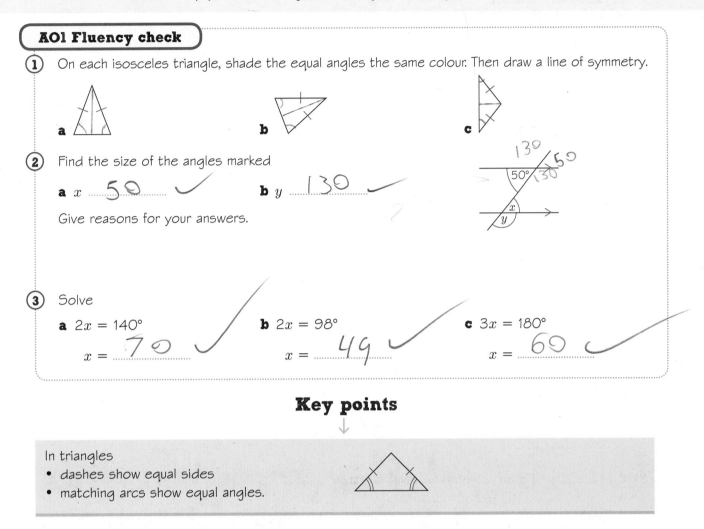

AO1 Fluency check

① On each isosceles triangle, shade the equal angles the same colour. Then draw a line of symmetry.

a

b

c

② Find the size of the angles marked

a x _____50_____ ✓ **b** y _____130_____ ✓

Give reasons for your answers.

130
50° 130
50
x
y

③ Solve

a $2x = 140°$ **b** $2x = 98°$ **c** $3x = 180°$

$x =$ ____70____ ✓ $x =$ ____49____ ✓ $x =$ ____60____ ✓

Key points
↓

In triangles
• dashes show equal sides
• matching arcs show equal angles.

These **skills boosts** will help you to find angles and lengths in triangles.

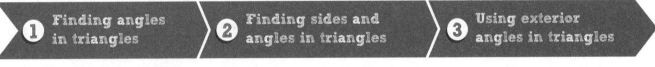

| ① Finding angles in triangles | ② Finding sides and angles in triangles | ③ Using exterior angles in triangles |

You might have already done some work on triangles. Before starting the first skills boost, rate your confidence with these questions.

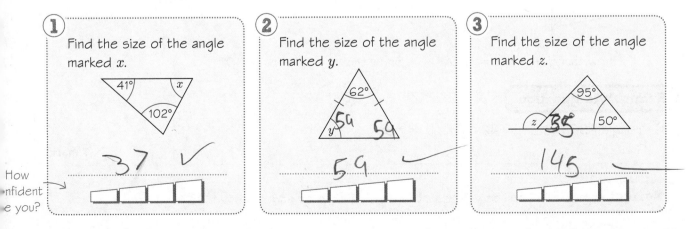

① Find the size of the angle marked x.

41° x
102°

____37____ ✓

② Find the size of the angle marked y.

62°
54 59
y

____59____ ✓

③ Find the size of the angle marked z.

95°
z 35° 50°

____145____ ✓

How confident are you?

1 Finding angles in triangles

The angles in a triangle add up to 180°.

Guided practice

Find angle x.
Give a reason for your answer.

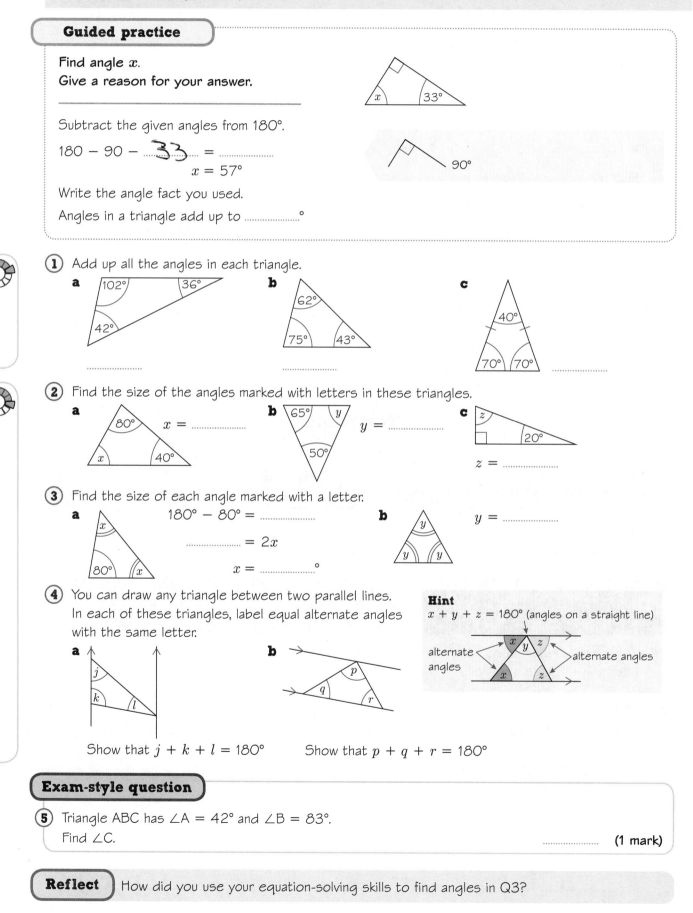

Subtract the given angles from 180°.

$180 - 90 - \underset{33}{\text{33}} = \underline{\hspace{2cm}}$

$x = 57°$

Write the angle fact you used.

Angles in a triangle add up to ....°

1 Add up all the angles in each triangle.

a 102° 36° 42°

b 62° 75° 43°

c 40° 70° 70°

....

2 Find the size of the angles marked with letters in these triangles.

a 80° $x =$ .... x 40°

b 65° y $y =$ .... 50°

c z 20° $z =$ ....

3 Find the size of each angle marked with a letter.

a x 80° x $180° - 80° = $ ....
.... $= 2x$
$x = $ ....°

b y y y $y = $ ....

4 You can draw any triangle between two parallel lines.
In each of these triangles, label equal alternate angles with the same letter.

a j k l

b p q r

Hint
$x + y + z = 180°$ (angles on a straight line)

alternate angles $\leftarrow$ x y z x z $\rightarrow$ alternate angles

Show that $j + k + l = 180°$ Show that $p + q + r = 180°$

Exam-style question

5 Triangle ABC has $\angle A = 42°$ and $\angle B = 83°$.
Find $\angle C$.   .... (1 mark)

Reflect
How did you use your equation-solving skills to find angles in Q3?

2 Finding sides and angles in triangles

An equilateral triangle has three equal sides and three equal angles.

An isosceles triangle has two equal sides and two equal angles at the bottom of those sides.

Guided practice

Find the size of the angle marked x.

_____

What type of triangle is it?   Equilateral or isosceles?

2 equal sides → .... triangle.

Label the equal angles.

Base angles in an isosceles triangle are equal.

$180 -$ .... $-$ .... $=$ ....

$x = 110°$ (angles in a triangle add up to 180°)

---

1 Label the missing sides and angles in the isosceles triangles.

a 65°   ......°   6cm   ......cm   50°

b 72°   10 cm   10 cm   54°   ......°

c 9 cm   72°   5 cm   ......°   72°   .... cm

2 Find the size of each angle marked with a letter.

a 55° q p ....

b r 4 cm s 20°   4 cm   ....

3 Triangle ABC has three equal angles.

a What type of triangle is it? ....

b Complete.

  i $x + x + x = 3x$

  $3x =$ ....°

  ii Length AC = ... cm

  Length BC = ... cm

A x 5 cm x x B   C

Exam-style question

4 In triangle XYZ, XY = YZ and ∠XYZ = 96°.

Find ∠ZXY.   .... (2 marks)

X   Y   96°   Z   Diagram NOT accurately drawn

Reflect   How did you use your equation-solving skills to find the angles in Q3 and Q4?

3 Using exterior angles in triangles

An exterior angle makes a straight line with an interior angle.

An exterior angle of a triangle is equal to the sum of the two opposite interior angles.

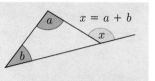

$x = a + b$

Guided practice

Find angles x and y.
Give reasons for your answers.

_____

Add the two angles inside the triangle on the opposite side.

$x = 50° +$ ....°
$x = 110°$

Write the angle fact you used.
Exterior angle equals sum of opposite interior angles.
Use angles on a straight line to find y.

$y =$ ....° $- 110°$

$y = 70°$

Angles on a ...., ...

Worked exam question

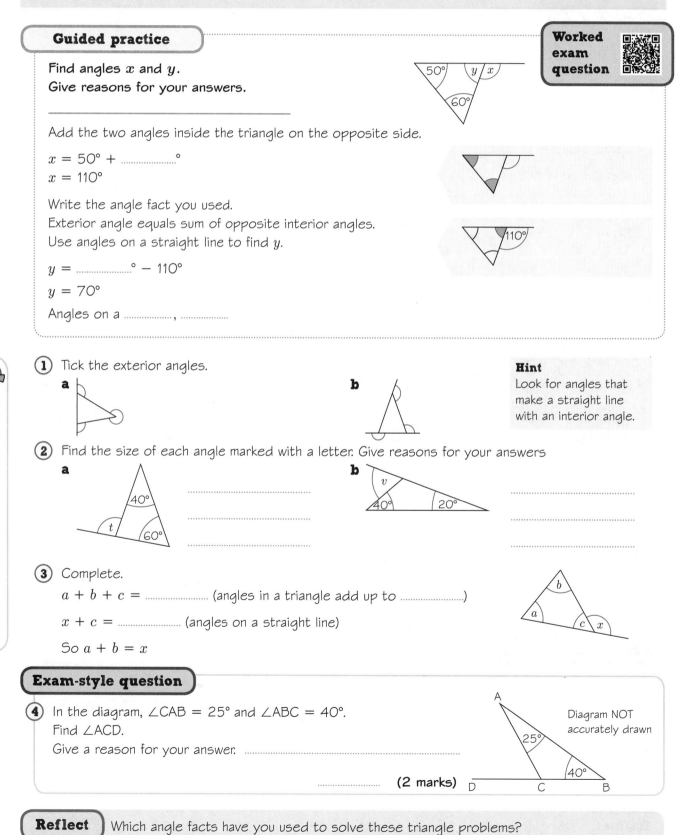

① Tick the exterior angles.

a

b

Hint
Look for angles that make a straight line with an interior angle.

② Find the size of each angle marked with a letter. Give reasons for your answers

a
40°
t 60°

....
....
....

b
v
40° 20°

....
....
....

③ Complete.

$a + b + c =$ .... (angles in a triangle add up to ....)

$x + c =$ .... (angles on a straight line)

So $a + b = x$

b
a c x

Exam-style question

④ In the diagram, $\angle CAB = 25°$ and $\angle ABC = 40°$.
Find $\angle ACD$.
Give a reason for your answer. ....

.... **(2 marks)**

A
25°
Diagram NOT accurately drawn
40°
D   C   B

Reflect Which angle facts have you used to solve these triangle problems?

Practise the methods

Answer this question to check where to start.

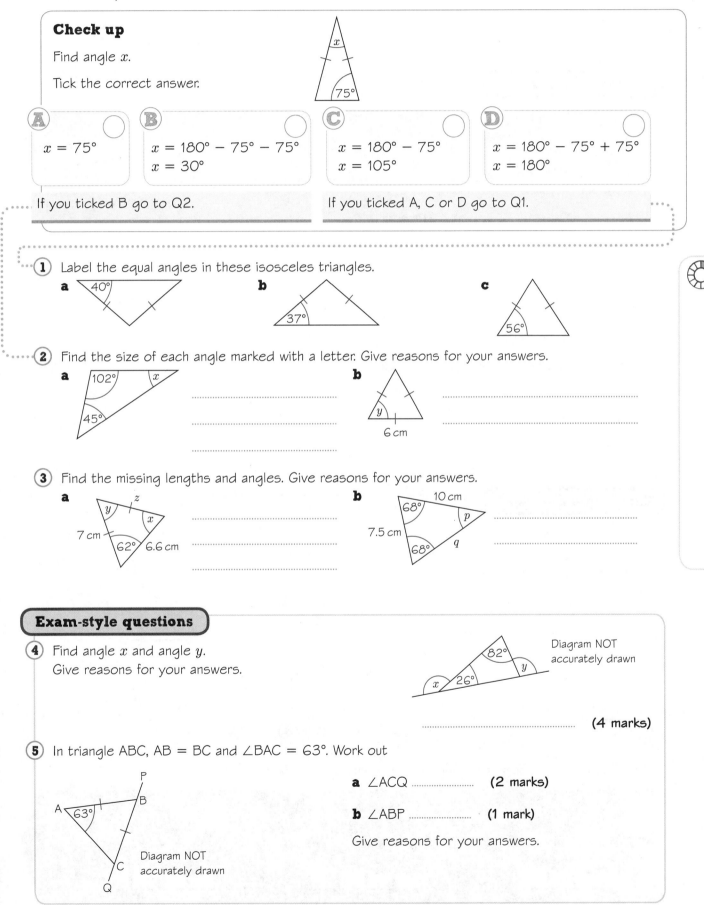

Check up

Find angle x.

Tick the correct answer.

75°

A ◯
$x = 75°$

B ◯
$x = 180° - 75° - 75°$
$x = 30°$

C ◯
$x = 180° - 75°$
$x = 105°$

D ◯
$x = 180° - 75° + 75°$
$x = 180°$

If you ticked B go to Q2.   If you ticked A, C or D go to Q1.

1 Label the equal angles in these isosceles triangles.

a 40°   **b** 37°   **c** 56°

2 Find the size of each angle marked with a letter. Give reasons for your answers.

a 102° x 45° ....

b y 6 cm ....

3 Find the missing lengths and angles. Give reasons for your answers.

a y z x 7 cm 62° 6.6 cm ....

b 68° 10 cm p 7.5 cm 68° q ....

Exam-style questions

4 Find angle x and angle y.
Give reasons for your answers.

82° y x 26°

Diagram NOT accurately drawn

.... (4 marks)

5 In triangle ABC, AB = BC and $\angle$BAC = 63°. Work out

P B A 63° C Q

Diagram NOT accurately drawn

a $\angle$ACQ .... (2 marks)

b $\angle$ABP .... (1 mark)

Give reasons for your answers.

Problem-solve!

(1) The triangle has sides of length 80 mm, 109 mm and 80 mm.

Label each side with its length.

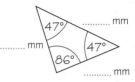

(2) Find the size of an exterior angle of an equilateral triangle.

.... **(2 marks)**

(3) In triangle ABC, AC = BC.
Work out the sizes of the angles marked

Diagram NOT accurately drawn

a x .... **(1 mark)**

b y .... **(2 marks)**

Give reasons for your answers.

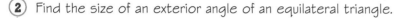

(4) ABCD is a kite. ∠DAB = 60°. AD = AB and DC = BC.

a Find the sizes of angles ADB and ABD. ....

b What type of triangle is ABD? ....

c Find the sizes of angles CDB and DBC. ....

d Hence, find the size of angle ADC. ....

(5) PQRS is a rhombus.

a Draw the two lines of symmetry. **(1 mark)**

b Work out the size of the angle marked

   i x .... **(1 mark)** **ii** y .... **(1 mark)**

Diagram NOT accurately drawn

(6) WXYZ is a square of side 5 cm.
V is a point on the diagonal WY such that WZ = WV. Find

a ∠ZWV .... **(1 mark)**

b ∠WZV .... **(1 mark)**

c ∠VZY .... **(1 mark)**

Diagram NOT accurately drawn

Now that you have completed this unit, how confident do you feel?

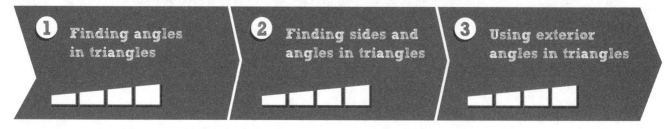

1 Finding angles in triangles

2 Finding sides and angles in triangles

3 Using exterior angles in triangles

⑥ Quadrilaterals

This unit will help you to find the area and perimeter of shapes made from rectangles, calculate angles and find missing vertices of quadrilaterals.

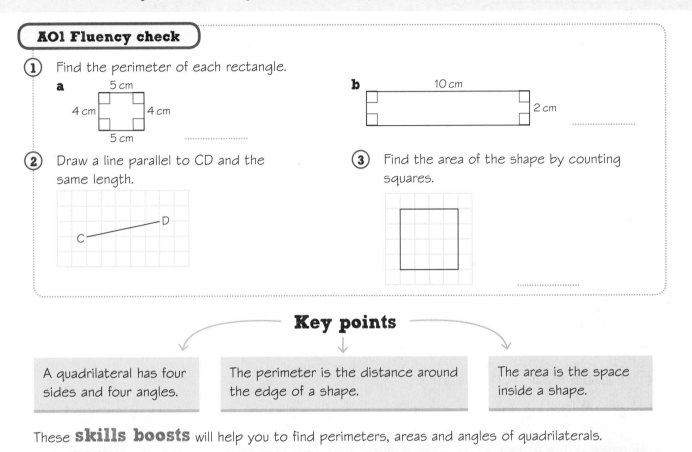

A01 Fluency check

① Find the perimeter of each rectangle.

a 5 cm / 4 cm / 4 cm / 5 cm

b 10 cm / 2 cm

② Draw a line parallel to CD and the same length.

C ——— D

③ Find the area of the shape by counting squares.

Key points

| A quadrilateral has four sides and four angles. | The perimeter is the distance around the edge of a shape. | The area is the space inside a shape. |

These **skills boosts** will help you to find perimeters, areas and angles of quadrilaterals.

1 Finding area and perimeter of shapes made from rectangles

2 Finding coordinates of missing vertices

3 Finding angles in quadrilaterals

You might have already done some work on quadrilaterals. Before starting the first skills boost, rate your confidence with these questions.

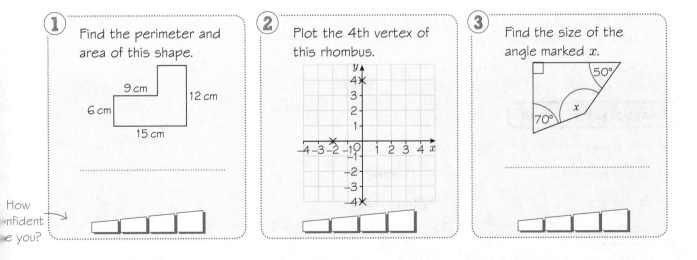

① Find the perimeter and area of this shape.

9 cm / 12 cm / 6 cm / 15 cm

② Plot the 4th vertex of this rhombus.

③ Find the size of the angle marked x.

50° / 70° / x

How confident are you?

1 Finding area and perimeter of shapes made from rectangles

Opposite sides of a rectangle are equal.

Area of a rectangle = length × width
Area is measured in mm², cm² or m².

Guided practice

For this shape, find

a the perimeter   **b** the area.

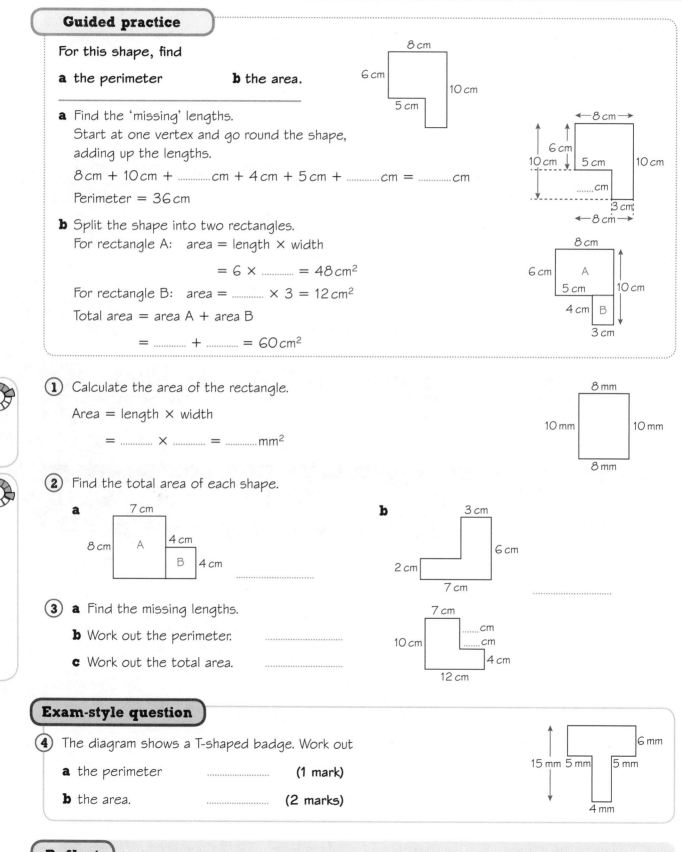

a Find the 'missing' lengths.
Start at one vertex and go round the shape,
adding up the lengths.

8 cm + 10 cm + ... cm + 4 cm + 5 cm + ... cm = ... cm

Perimeter = 36 cm

b Split the shape into two rectangles.
For rectangle A: area = length × width

= 6 × .... = 48 cm²

For rectangle B: area = ... × 3 = 12 cm²

Total area = area A + area B

= .... + ... = 60 cm²

(1) Calculate the area of the rectangle.

Area = length × width

= .... × .... = ... mm²

(2) Find the total area of each shape.

a

b

(3) a Find the missing lengths.

b Work out the perimeter.   ....

c Work out the total area.   ....

Exam-style question

(4) The diagram shows a T-shaped badge. Work out

a the perimeter   ....   (1 mark)

b the area.   ....   (2 marks)

Reflect   In these questions, when did you use addition and when did you use subtraction?

2 Finding coordinates of missing vertices

Parallelogram: two pairs of parallel sides, opposite sides are equal.   Rhombus: two pairs of parallel sides, all sides are equal.

A (1, 1), B (4, 1) and D (−1, 4) are three vertices of a parallelogram ABCD.
Find the coordinates of the 4th vertex, C.

Plot and label the points A, B and D.

Join the points.

Draw one side parallel and equal to AB.   AB = DC and
Draw the other side parallel and equal to AD.   AD = BC
Write the coordinates of C.

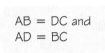

C = (2, ...)

---

① E (2, 1), F (4, 1) and G (4, 3) are three vertices of a square EFGH.

Find the coordinates of the 4th vertex, H.

H = (............, ...)

② J (−4, 2), K (−2, 2) and M (−4, −3) are three vertices of a rectangle JKLM.

Find the coordinates of the 4th vertex, L.

L = (............, ...)

③ P (3, 3), Q (1, 2) and R (3, −3) are three vertices of a kite PQRS.

Find the coordinates of the 4th vertex, S.

S = (............, ...)

Use this diagram for Questions 1 to 3

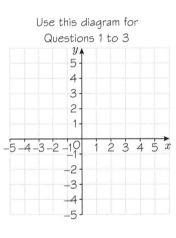

④ A rhombus WXYZ has vertices W (−5, 0), X (0, 3) and Y (5, 0).

Find the coordinates of the 4th vertex, Z.

.... **(2 marks)**

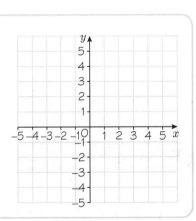

Reflect   How did you use the properties of quadrilaterals (for example, facts about equal sides or parallel sides) to answer these questions?

3 Finding angles in quadrilaterals

The angles in a triangle add up to 180°. The angles in a quadrilateral add up to 360°.

Guided practice

Find angle x.

Give a reason for your answer.

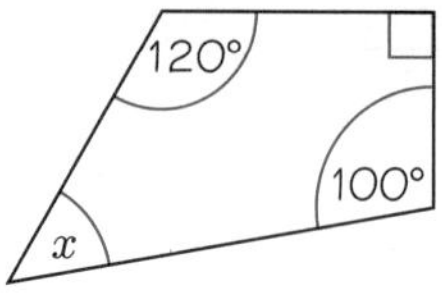

_____

Subtract the given angles from 360°.

$360 - 120 - \text{..........} - 90 = \text{..........}$

$x = 50°$

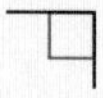

Write the angle fact you used.
Angles in a quadrilateral add up to 360°.

90°

1. Add up the angles in the square and in the rectangle.

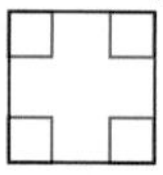

   Angles add up to 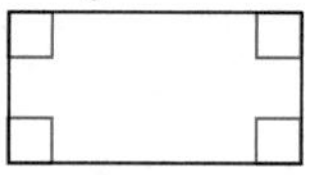 ....°   Angles add up to ....°

2. You can divide this quadrilateral into two triangles.

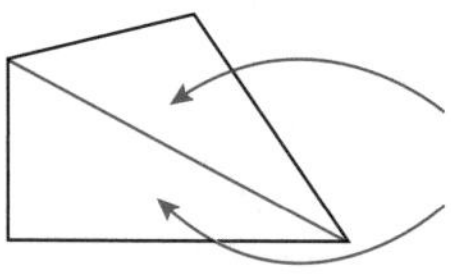

   angle sum of triangle = 180°
   angle sum of triangle = 180° } angle sum of quadrilateral = 180° + 180° = 360°

   Show that you can divide these quadrilaterals into two triangles, so each angle sum is 360°.

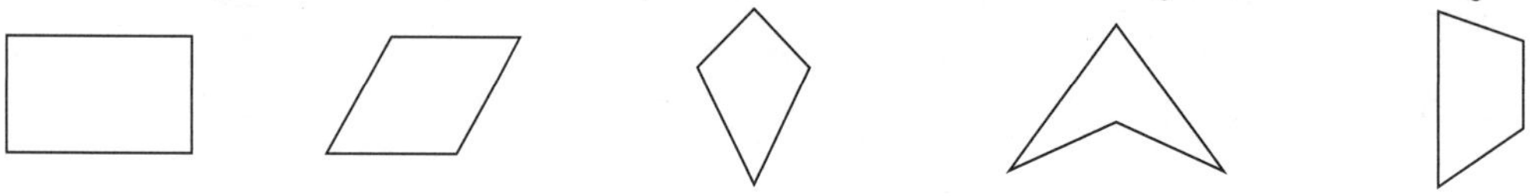

3. Work out the size of each angle marked with a letter.
   a 130° x 50° 130° ....
   b y 120° 80° ....
   c 125° 70° z 125° ....

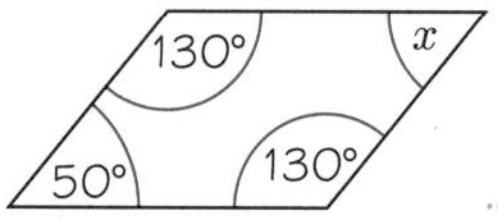

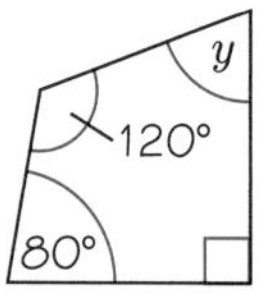

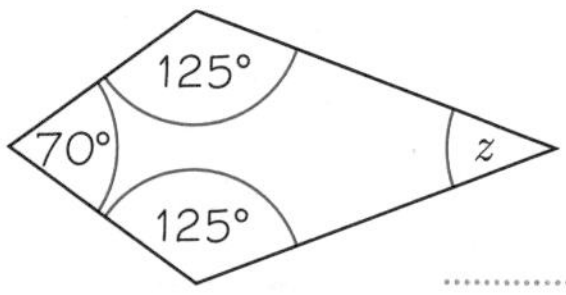

4. Work out the size of each angle marked with a letter.
   a x 80°
      Hint
      Opposite angles of a parallelogram are equal.
      $360° - 80° - \text{..........}° = 2x$
      $x = \text{..........}°$
   b 105° b c a ....

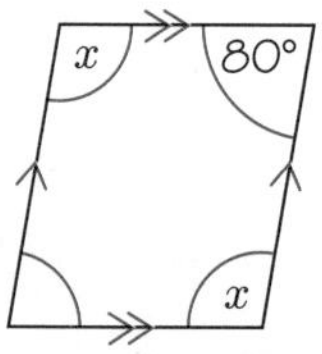

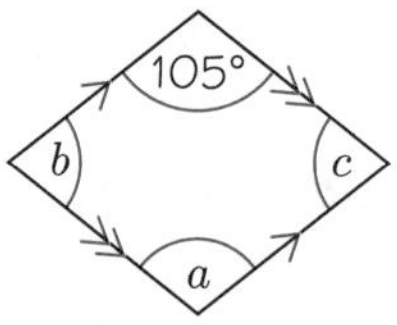

Exam-style question

5. In quadrilateral ABCD, ∠A = 36°, ∠B = 90° and ∠C = 114°.

   Find ∠D. .... (2 marks)

Reflect   How did you use your equation-solving skills to answer Q4?

Practise the methods

Answer this question to check where to start.

Check up

Find the area and perimeter of this shape.

Tick the correct answer.

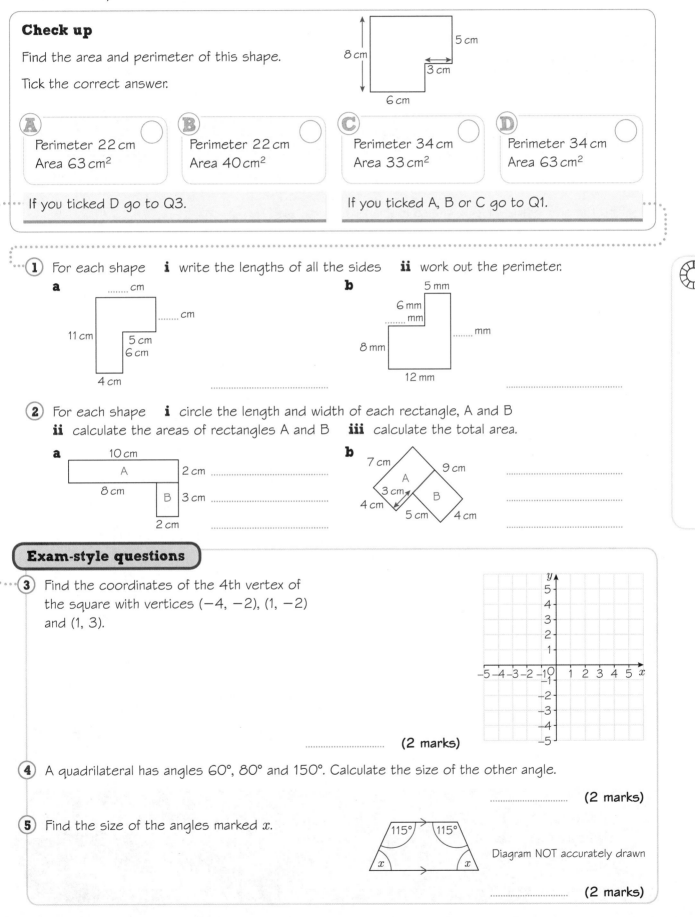

A	B	C	D
Perimeter 22 cm Area 63 cm² ○	Perimeter 22 cm Area 40 cm² ○	Perimeter 34 cm Area 33 cm² ○	Perimeter 34 cm Area 63 cm² ○

If you ticked D go to Q3.   If you ticked A, B or C go to Q1.

1 For each shape **i** write the lengths of all the sides **ii** work out the perimeter.

a

b

2 For each shape **i** circle the length and width of each rectangle, A and B
ii calculate the areas of rectangles A and B **iii** calculate the total area.

a

b

Exam-style questions

3 Find the coordinates of the 4th vertex of the square with vertices (−4, −2), (1, −2) and (1, 3).

.... (2 marks)

4 A quadrilateral has angles 60°, 80° and 150°. Calculate the size of the other angle.

.... (2 marks)

5 Find the size of the angles marked x.

Diagram NOT accurately drawn

.... (2 marks)

Unit 6 Quadrilaterals   39

Problem-solve!

(1) Find the perimeter and area of a square of side 6 cm.

....

Exam-style questions

(2) A square of side 5 cm is cut from the corner of a sheet of card 20 cm by 30 cm.

Calculate the area of the remaining card.

5 cm

5 cm

20 cm

30 cm

.... **(2 marks)**

(3) Find the size of the missing angles in this kite.

30°

Diagram NOT accurately drawn

.... **(2 marks)**

(4) The diagram shows the plan of a garden.

Calculate the area and perimeter of the lawn.

20 m

3 m

Patio

4 m   lawn   8 m

.... **(3 marks)**

(5) A, B and C are three vertices of a parallelogram.

a Join A to B and B to C.

Hint

A   D

B   C

Find the coordinates of the 4th vertex, D.

....

b Join A to C and A to B.

Hint

A

B   C

E

Find the coordinates of the 4th vertex, E.

....

Exam-style question

(6) Find the angle sum of this pentagon.

.... **(2 marks)**

Now that you have completed this unit, how confident do you feel?

1 Finding area and perimeter of shapes made from rectangles

2 Finding coordinates of missing vertices

3 Finding angles in quadrilaterals

⑦ Angles

This unit will help you to find missing angles in crossing and parallel lines.

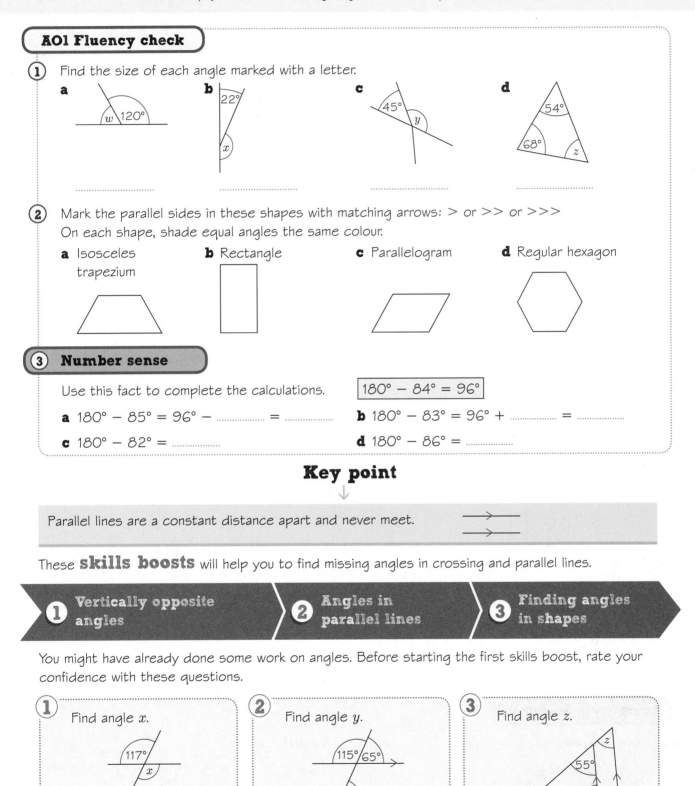

AO1 Fluency check

① Find the size of each angle marked with a letter.

a w \\ 120° **b** 22° x **c** 45° y **d** 54° 68° z

② Mark the parallel sides in these shapes with matching arrows: > or >> or >>>
On each shape, shade equal angles the same colour.

a Isosceles trapezium   **b** Rectangle   **c** Parallelogram   **d** Regular hexagon

③ Number sense

Use this fact to complete the calculations. $180° − 84° = 96°$

a $180° − 85° = 96° − \text{.....} = \text{.....}$ **b** $180° − 83° = 96° + \text{.....} = \text{.....}$

c $180° − 82° = \text{.....}$ **d** $180° − 86° = \text{.....}$

Key point
↓

Parallel lines are a constant distance apart and never meet.

These **skills boosts** will help you to find missing angles in crossing and parallel lines.

① Vertically opposite angles   **②** Angles in parallel lines   **③** Finding angles in shapes

You might have already done some work on angles. Before starting the first skills boost, rate your confidence with these questions.

① Find angle x. 117° x

② Find angle y. 115° 65° y

③ Find angle z. z 55° 40° 85°

How confident are you?

1 Vertically opposite angles

Two lines that cross make two pairs of vertically opposite angles.
'Vertically opposite' means their vertices (points) are opposite each other
(not that the angles are vertical).
Vertically opposite angles are equal.

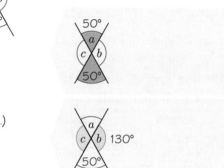

Guided practice

Find the angles marked with letters.
Give reasons for your answers.

Look for vertically opposite angles.
Write the reason.

$a = 50°$ (.............................. opposite angles)

Look for angles on a straight line.

$b + 50° = $ ....° (angles on a ....)

$\qquad b = 130°$

$c = 130°$ (.............................. opposite angles)

① Find the size of each angle marked with a letter.

a

$x = $ ....°

b

$y = $ ....°

c

$z = $ ....°

② Find the angles labelled with letters.

a ....

....

b ....

....

c ....

....

③ Find the size of each angle marked with a letter.
Give reasons for your answers.

a ....

....

....

b ....

....

....

Exam-style question

④ Find **a** angle x .... (1 mark)

b angle y .... (1 mark)

Give reasons for your answers.

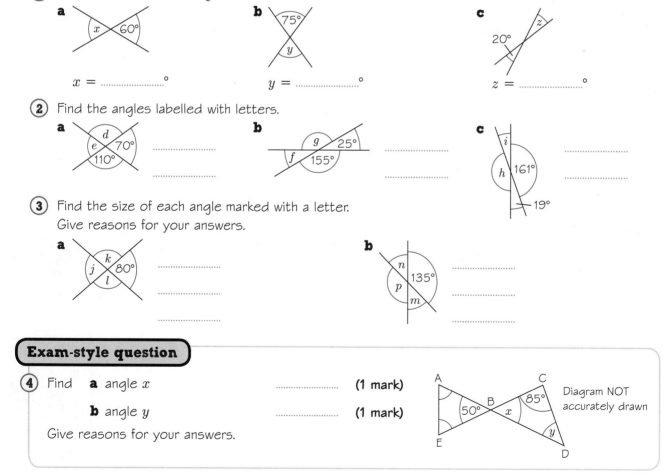

Diagram NOT
accurately drawn

Reflect Look back at Q3a. Find k using angles on a straight line. Then find j using angles on a
straight line. Can you find l in the same way?

2 Angles in parallel lines

In parallel lines
• Corresponding angles are equal.
• Alternate angles are equal.

Guided practice

Find the size of each angle marked with a letter.
Give reasons for your answers.

_____

Are a and 70° corresponding or alternate angles?

$a = 70°$ (............................... angles)

Are a and b corresponding or alternate angles?

$b = $ ....° (corresponding angles)

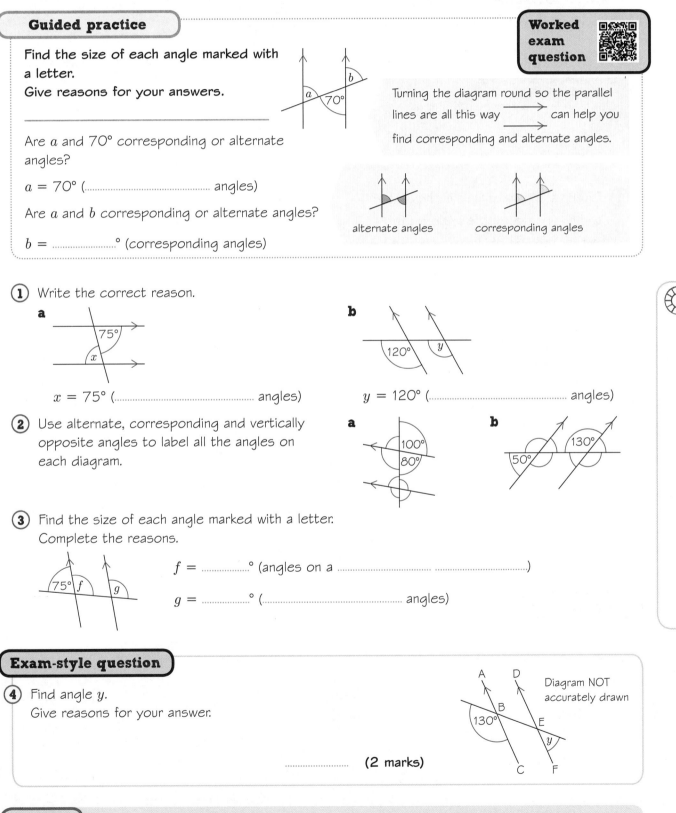

Worked exam question

Turning the diagram round so the parallel lines are all this way ⟶ can help you find corresponding and alternate angles.

alternate angles   corresponding angles

① Write the correct reason.

a

$x = 75°$ (............................... angles)

b

$y = 120°$ (............................... angles)

② Use alternate, corresponding and vertically opposite angles to label all the angles on each diagram.

a 100° 80°

b 50° 130°

③ Find the size of each angle marked with a letter. Complete the reasons.

75° f g

$f = $ ....° (angles on a ....)

$g = $ ....° (............................... angles)

Exam-style question

④ Find angle y.
Give reasons for your answer.

A D   Diagram NOT accurately drawn

B

130°   E

y

C   F

.... **(2 marks)**

Reflect   Draw several diagrams with two parallel lines and a crossing line. Measure the corresponding angles and confirm that they are equal. Repeat for alternate angles.

3 Finding angles in shapes

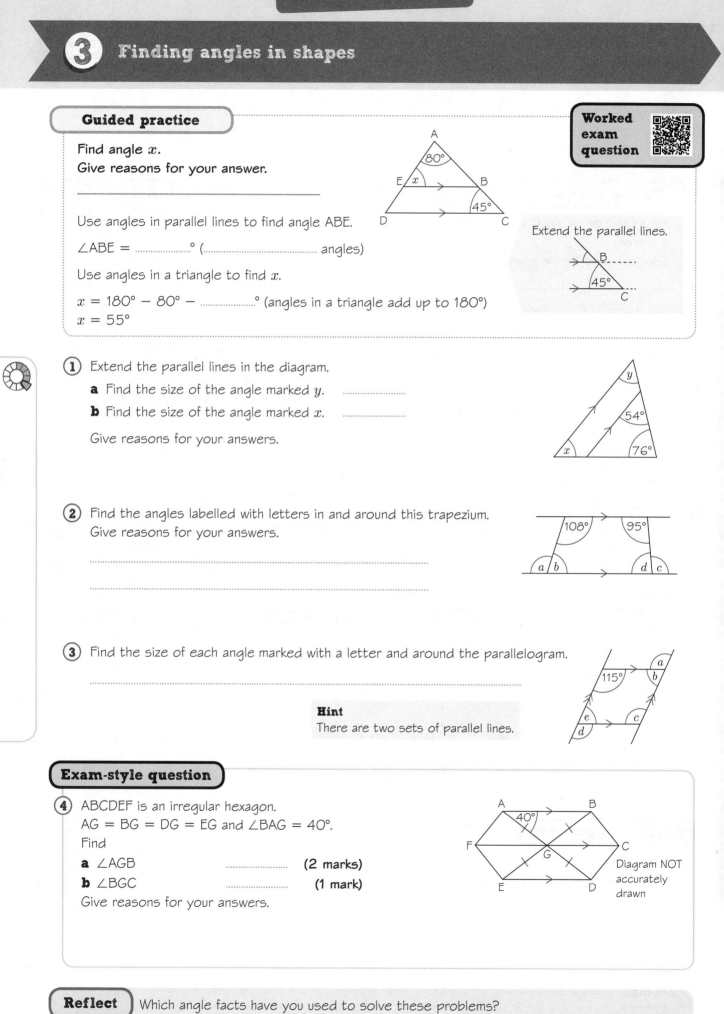

Guided practice

Find angle x.
Give reasons for your answer.

Use angles in parallel lines to find angle ABE.

∠ABE = ...° (................ angles)

Use angles in a triangle to find x.

$x = 180° − 80° − $ ....° (angles in a triangle add up to 180°)
$x = 55°$

Worked exam question

Extend the parallel lines.

① Extend the parallel lines in the diagram.
  a Find the size of the angle marked y. ....
  b Find the size of the angle marked x. ....

  Give reasons for your answers.

② Find the angles labelled with letters in and around this trapezium.
  Give reasons for your answers.

③ Find the size of each angle marked with a letter and around the parallelogram.

Hint
There are two sets of parallel lines.

Exam-style question

④ ABCDEF is an irregular hexagon.
  AG = BG = DG = EG and ∠BAG = 40°.
  Find
  a ∠AGB .... (2 marks)
  b ∠BGC .... (1 mark)
  Give reasons for your answers.

Diagram NOT accurately drawn

Reflect   Which angle facts have you used to solve these problems?

Practise the methods

Answer this question to check where to start.

Check up

Look at the angles in the diagram.

Tick the statement that is correct.

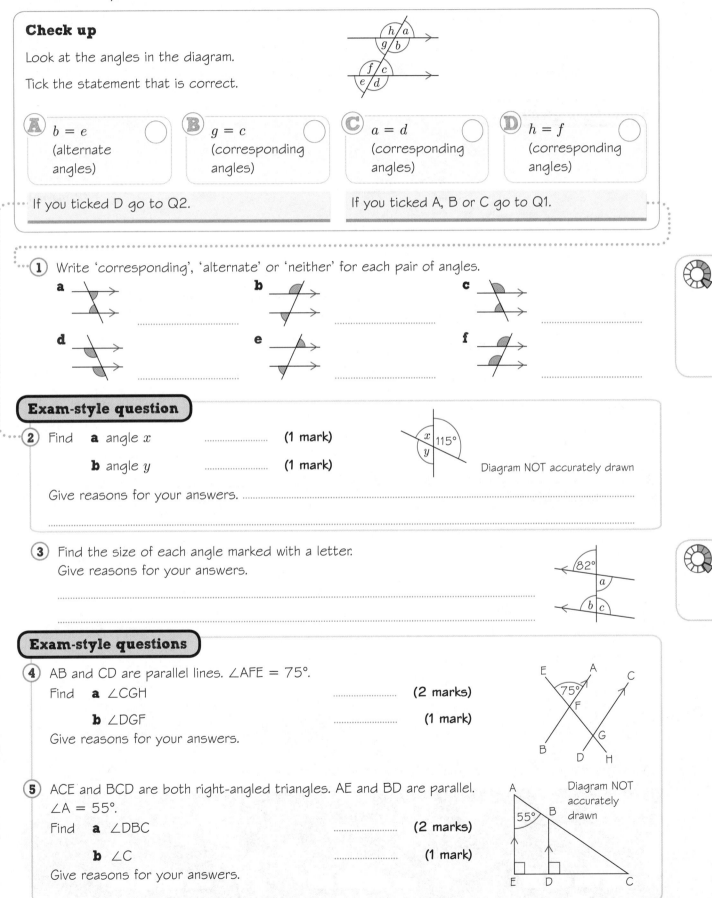

A $b = e$ ◯
(alternate
angles)

B $g = c$ ◯
(corresponding
angles)

C $a = d$ ◯
(corresponding
angles)

D $h = f$ ◯
(corresponding
angles)

If you ticked D go to Q2.   If you ticked A, B or C go to Q1.

1 Write 'corresponding', 'alternate' or 'neither' for each pair of angles.

a   **b**   **c**

....

d   **e**   **f**

....

Exam-style question

2 Find **a** angle x .... (1 mark)

b angle y .... (1 mark)

x 115°
y

Diagram NOT accurately drawn

Give reasons for your answers. ....

....

3 Find the size of each angle marked with a letter.
Give reasons for your answers.

82°
a
b c

....

....

Exam-style questions

4 AB and CD are parallel lines. ∠AFE = 75°.
Find **a** ∠CGH   .... (2 marks)

b ∠DGF   .... (1 mark)

Give reasons for your answers.

E   A
C
75°
F
B
G
D   H

5 ACE and BCD are both right-angled triangles. AE and BD are parallel.
∠A = 55°.
Find **a** ∠DBC   .... (2 marks)

b ∠C   .... (1 mark)

Give reasons for your answers.

A   Diagram NOT
accurately
55° B   drawn

E   D   C

Problem-solve!

① Find the size of each angle marked with a letter.

Give reasons for your answers.

....

....

② Find the value of x.

$x + 40°$

$2x - 50°$

Diagram NOT accurately drawn

.... **(2 marks)**

③ ABCD is a rectangle.
Diagonals AC and BD intersect at E.
AE = DE = BE = CE and ∠DAE = 62°

Find

Diagram NOT accurately drawn

a ∠ADE .... **(1 mark)**

b ∠AED .... **(1 mark)**

c ∠BEC .... **(1 mark)**

Give reasons for your answers.

④ Show that triangle ABC has the same size angles
as triangle DEC.

(5 marks)

⑤ DEFG is a parallelogram.

Use angle facts for parallel lines to label all
the angles that are equal to d.

Now that you have completed this unit, how confident do you feel?

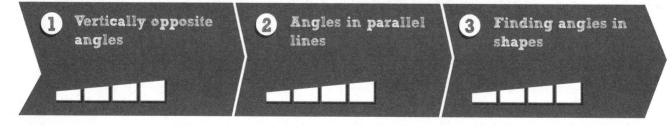

1 Vertically opposite angles

2 Angles in parallel lines

3 Finding angles in shapes

 Loci

This unit will help you to draw accurate constructions and loci.

(1) Construct a circle, with centre O and radius 2.5 cm.

O •

(2) Draw a line exactly 6 cm long.

(3) Describe a regular hexagon.

....

....

Key points

| A locus is the path that an object follows. | Loci is the plural of locus. | 'Construct' a locus or shape means 'draw accurately using only compasses and a ruler'. |

These **skills boosts** will help you to draw constructions.

| 1 Constructing triangles and hexagons | 2 Constructing loci equidistant from points | 3 Constructing loci around lines and corners |

You might have already done some work on loci. Before starting the first skills boost, rate your confidence using these questions. Draw your constructions on plain paper.

(1) Construct an equilateral triangle of side 4 cm.

(2) Construct the locus of points 4 cm from a point.

(3) Construct the locus of points 2 cm from a straight line.

How confident are you?

1 Constructing triangles and hexagons

Guided practice

Construct an equilateral triangle of side 5 cm.

Follow the instructions to construct the triangle on the line.

Draw a line 5 cm long.

Use a ruler.
Measure it accurately.

Open your compasses to 5 cm.

Put the compass point on one end of the line. Draw an arc.

Put the compass point on the other end of the line.

Draw an arc to cross the first one.

Join the point where the arcs cross to the ends of the line. Use a ruler.

Do not rub out the arcs.

(1) Construct an equilateral triangle of side 3 cm.

Hint
Draw a line 3 cm long.
Open your compasses to 3 cm.

2 Follow the instructions to construct a regular hexagon.

Open your compasses to the radius of the circle.

Hint
Put the point on the
centre and pencil on
the circumference.

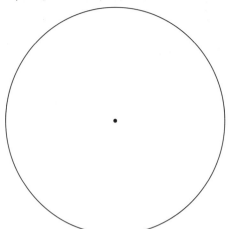

Put your compass point anywhere on
the circumference.

Draw an arc on the circumference.

Move the compass point to where your arc
crosses the circumference.

Draw another arc.
Continue round the circle.

Join up your arcs with straight lines.

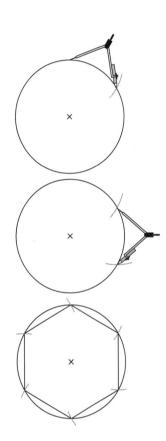

Exam-style question

3 Construct an equilateral triangle of side 6 cm.

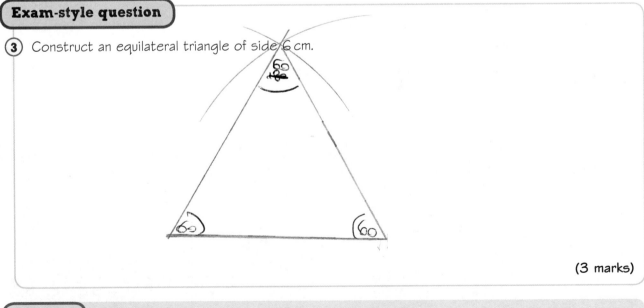

(3 marks)

Reflect Measure the sides of your triangles and hexagon.
All the lengths should be within 1 to 2 mm of each other.

2 Constructing loci equidistant from points

The locus of points equidistant from a fixed point is a circle.
The locus of points equidistant from points X and Y is the perpendicular bisector of XY.

Guided practice

Construct the locus of a point that is equidistant from points A and B.

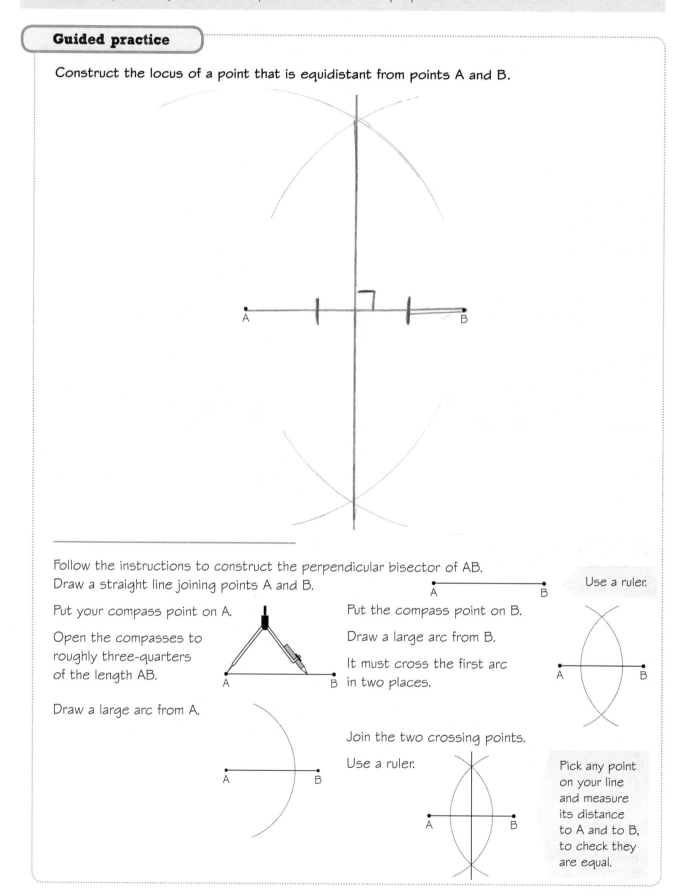

Follow the instructions to construct the perpendicular bisector of AB.
Draw a straight line joining points A and B.

Use a ruler.

Put your compass point on A.

Open the compasses to roughly three-quarters of the length AB.

Draw a large arc from A.

Put the compass point on B.

Draw a large arc from B.

It must cross the first arc in two places.

Join the two crossing points.

Use a ruler.

Pick any point on your line and measure its distance to A and to B, to check they are equal.

① Construct the locus of points 3 cm from C.

•
C

Exam-style question

② A road is to be built equidistant from two towns, Maidford and Plumpton.
Construct an accurate diagram to show the position of the road.

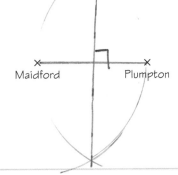

Maidford   Plumpton

(3 marks)

③ A goat in a field is tethered to a point P by a rope 4 m long.
Shade the region the goat can reach.

Scale: 1 cm represents 1 m

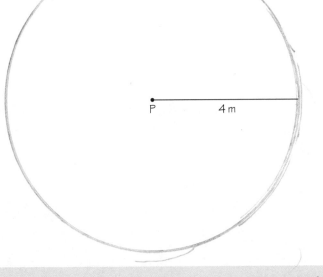

P   4 m

Reflect How does using a sharp pencil improve the accuracy of constructions?

3 Constructing loci around lines and corners

Guided practice

Draw accurately the locus of points 2 cm from the line.

Follow the instructions to draw the locus around the line.

Draw two lines, each 2 cm from
the line and parallel to it.

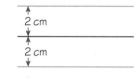

Open your compasses to 2 cm.

Draw a semicircle radius 2 cm
centred on each end of the line.

Every point on the
semicircle is 2 cm from
the end of the line.

① Draw a line 6 cm long.
Construct the locus of points 3 cm from the line.

(2) Follow the instructions to draw accurately the locus of points 1.5 cm outside the rectangle.

Draw lines 1.5 cm from and parallel to each side.

Open your compasses to 1.5 cm.

Draw an arc of radius 1.5 cm centred on each corner.

Exam-style question

(3) Draw accurately the locus of points 2 cm outside the square.

(3 marks)

Reflect How have you used 'the locus of points equidistant from a fixed point is a circle' to construct loci round corners?

Practise the methods

Answer this question to check where to start.

Check up

Draw accurately the locus of points 1 cm from the line XY.

Tick the drawing that is correct.

X _____ Y

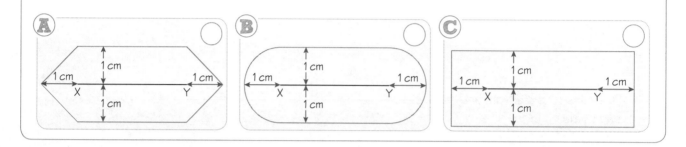

A ○   **B** ○   **C** ○

With diagrams showing:
- A: hexagon shape with 1 cm markings around X and Y
- B: rounded rectangle (stadium) shape with 1 cm markings around X and Y
- C: rectangle with 1 cm markings around X and Y

If you ticked B go to Q2.   If you ticked A or C go to Q1.

1 **a** In diagram A, measure the distance from X to any point on the sloping line.

Is it 1 cm? ....

b In diagram C, measure the distance from X to the corner.

Is it 1 cm? ....

c In diagram B, measure the distance from X to any point on the semicircle.

Is it 1 cm? ....

d Draw accurately the locus of points 1 cm from the line PQ.

P _____ Q

2 On plain paper, construct a regular hexagon inside a circle of radius 4 cm.

3 Construct the locus of points equidistant from M and N.

• M

• N

Exam-style question

4 A horse is tied to the corner inside a stable by a rope 2 metres long.

Construct and shade the region where the horse can move.

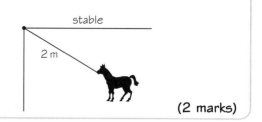

stable

2 m

(2 marks)

Problem-solve!

1. **a** Construct an equilateral triangle
   of side 4 cm. **(3 marks)**
   b Draw accurately the locus of
   points 1.5 cm outside the triangle. **(3 marks)**

2. In the diagram, X and Y represent two mobile phone masts.
   The signal from X reaches up to 8 km.
   The signal from Y reaches up to 10 km.
   Using a scale of 1 cm to 2 km, construct the regions the signals cover.
   Shade the region covered by signals from both masts.

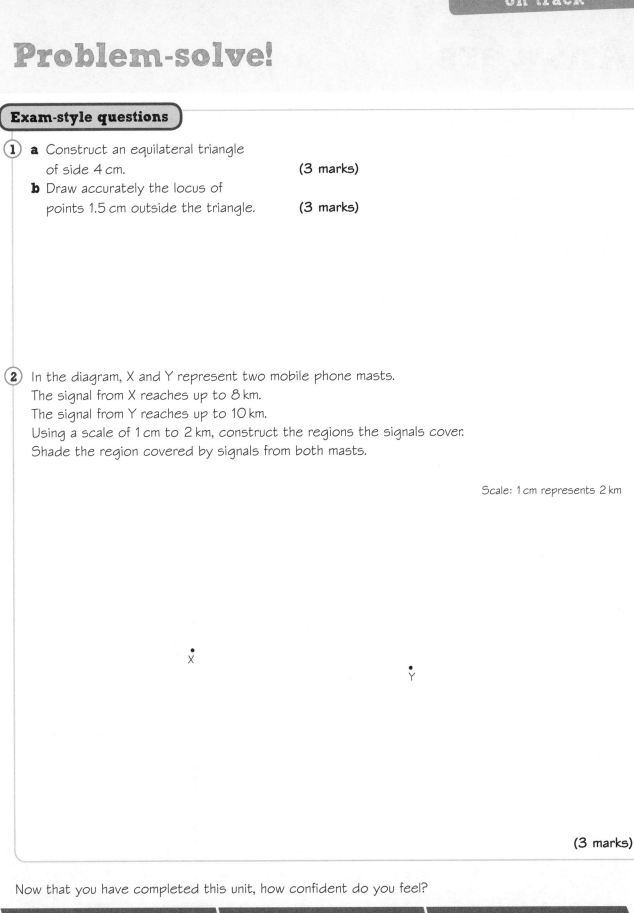

Scale: 1 cm represents 2 km

(3 marks)

Now that you have completed this unit, how confident do you feel?

1 Constructing triangles and hexagons

2 Constructing loci equidistant from points

3 Constructing loci around lines and corners

Answers

Unit 1 Simplifying and brackets

A01 Fluency check

① **a** 4   **b** 4   **c** -3
   d -8 **e** -15 **f** 12
② **a** 3   **b** 1   **c** -1
   d -3 **e** -3 **f** -1
③ **a** $6a$ **b** $2a$ **c** $5a$

④ Number sense

A iii, B i, C ii

Confidence questions

① $3x + 4y$
② $12xy$
③ $5x - 30$
④ $3(2x - 5)$

Skills boost 1 Simplifying by collecting like terms

Guided practice

a $a + 2b + 5a$
$= \circled{a} + 2b + \circled{5a}$
$= \underline{6} \, a + 2b$
b $5x + 2y - 3x + y$
$= \circled{5x} + \circled{2y} - \circled{3x} + \circled{y}$
$= \underline{5} \, x - \underline{3} \, x + 2y + y$
$= 2x + \underline{3} \, y$
① $7x + 5$
② **a** $7m + 3$ **b** $6y + 2$
   c $8t + 2$ **d** $9a + 3b$
   e $3m + 5p$ **f** $7x + 3y$
③ **a** $x + 3$ **b** $4a - 5b$
   c $2n^2 - 2n$ **d** $-4m + 4t + 6$
④ **a** $2x - 2y + 5$ **b** $3z^2 + 7z$

Skills boost 2 Multiplying terms

Guided practice

a $4a \times 2b$
$= \underline{8} \, ab$
b $c \times 2c$
$= \underline{2} \times c \cdot \underline{2}$
$= 2c^2$
① **a** $6a$ **b** $12t$ **c** $10x$ **d** $6n$
② **a** cd **b** $5pr$ **c** $4xy$
③ **a** $15ct$ **b** $12xy$ **c** $12fg$
④ **a** $-8a$ **b** $-6c$ **c** $-6xy$
   d $-10nt$ **e** $8sy$ **f** $-2nr$
⑤ **a** $8a^2$ **b** $3m^2$ **c** $-2x^2$
   d $-6y^2$ **e** $-4z^2$ **f** $6b^2$
⑥ **a** $6abc$ **b** $-10d^2$

Skills boost 3 Expanding brackets

Guided practice

a $2(x + 4)$
$2 \times x = \underline{2x}$
$= 2(x + 4)$
$2 \times \underline{4}$
$= 2x + \underline{8}$
b $3(x - 2)$
$3 \times x = \underline{3x}$
$= 3(x - 2)$
-6
$= 3x - \underline{6}$
① **a** $3x + 3$ **b** $4n + 12$
② **a** $3x + 6$ **b** $4y + 4$ **c** $3t - 9$
   d $2x + 10$ **e** $5m + 15$ **f** $2x + 12$
③ **a** $5x - 5$ **b** $4y - 8$ **c** $5z + 20$
   d $2x - 10$ **e** $7n - 14$ **f** $10z - 50$
④ **a** $6a + 8$ **b** $12x + 3$ **c** $10m + 4$
   d $6x - 3$ **e** $8a - 12$ **f** $15x - 10$
⑤ **a** $2p + 14$ **b** $12a - 6$

Skills boost 4 Factorising expressions

Guided practice

$3 \times \boxed{x} = 6$
$3x + 6 = 3(\boxed{x} + \boxed{2})$
$3 \times \boxed{2} = 6$
① **a** $5(x + 2)$ **b** $2(x + 3)$
② **a** $3(x + 3)$ **b** $4(d + 3)$ **c** $2(x + 4)$
③ **a** $5(x - 3)$ **b** $2(x - 2)$ **c** $3(f - 1)$
④ **a** $2(x + 1)$ **b** $3(a - 1)$ **c** $2(3n + 4)$
⑤ **a** $5(m + 4)$ **b** $3(3x - 5)$

Practise the methods

① **a** $2x$ **b** $4y$ **c** $5a + 2b$ **d** $3m + 4t$
② **a** $5ab$ **b** $6mp$ **c** $8xy$ **d** $21xy$
③ **a** $5a + 20$ **b** $7x + 14$ **c** $4y - 8$
④ **a** $3(x + 10)$ **b** $4(x - 5)$ **c** $3(3x + 4)$
   d $4(2y + 5)$ **e** $5(2z - 3)$ **f** $2(3x - 4)$
⑤ **a** $11x - 2y$ **b** $15x^2$
⑥ **a** $6x + 18$ **b** $10m - 5$
⑦ **a** $12x + 15$ **b** $7(y - 5)$

Problem-solve!

① **a** $2x + 2 + 9y$ **b** $2f - 7g + 2$
② LHS $= 7x + 2x - 3y + 4y + 5 - 8$
      $= 9x + y - 3 =$ RHS
③ **a** $3 + 3x$ **b** $20 - 5x$ **c** $22 + 10x$
④ **a** length $\times$ width $= 5 \times (x + 2) = 5(x + 2)$
   b $5(x + 2) = 5x + 10$
⑤ $3(x - 4)$ or $3x - 12$
⑥ **a** $2x + 2y$ **b** $6x + 3z$ **c** $20m - 5n$
   d $12t - 20w$ **e** $12s - 18x$ **f** $20x - 8y$

(7) **A ii, B iii, C i**
(8) **a** $2(2x + 7)$ **b** $x^2 + 2x$

Unit 2 Equations, expressions and formulae

AO1 Fluency check

(1) **a** 9   **b** 6   **c** 49   **d** 14
(2) **a** $2x$ **b** $2x$ **c** $4x^2 + 2x$
(3) **a** $3w + 6$ **b** $5x - 15$ **c** $8x + 4$
(4) **a**

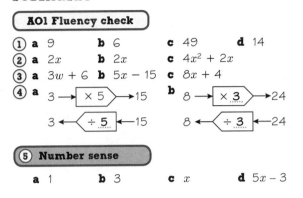

(5) **Number sense**

   a 1   **b** 3   **c** x **d** $5x - 3$

Confidence questions
(1) $x = 5$ (2) $x = 3$ (3) $30n$ (4) $R = 12$

Skills boost 1 Solving equations with the letter on one side

Guided practice

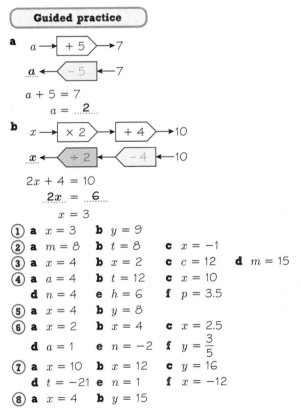

a

$a + 5 = 7$
$a = \underline{2}$

b

$2x + 4 = 10$
$\underline{2x} = \underline{6}$
$x = 3$

(1) **a** $x = 3$ **b** $y = 9$
(2) **a** $m = 8$ **b** $t = 8$ **c** $x = -1$
(3) **a** $x = 4$ **b** $x = 2$ **c** $c = 12$ **d** $m = 15$
(4) **a** $a = 4$ **b** $t = 12$ **c** $x = 10$
   d $n = 4$ **e** $h = 6$ **f** $p = 3.5$
(5) **a** $x = 4$ **b** $y = 8$
(6) **a** $x = 2$ **b** $x = 4$ **c** $x = 2.5$
   d $a = 1$ **e** $n = -2$ **f** $y = \frac{3}{5}$
(7) **a** $x = 10$ **b** $x = 12$ **c** $y = 16$
   d $t = -21$ **e** $n = 1$ **f** $x = -12$
(8) **a** $x = 4$ **b** $y = 15$

Skills boost 2 Solving equations with the letter on both sides

Guided practice

$3x + 5 = 6x + 2$
$5 = \underline{3x} + 2$
$\underline{3} = 3x$
$1 = x$

(1) **a** $x = 4$ **b** $x = 6$
(2) **a** $x = 2$ **b** $y = 5$
(3) **a** $x = 1$ **b** $y = 2$
(4) **a** $w = 8$ **b** $x = 6$
(5) $x = 4$

Skills boost 3 Writing expressions

Guided practice

a 1 box x balls
   $\underline{3}$ boxes $3x$ balls
b 1 box x balls
   2 boxes $\underline{2x}$ balls
   $2x + \underline{3}$

(1) **a** $n + 3$ **b** $y + 4$ **c** $5m$
   d $3x$ **e** $z - 3$ **f** $\frac{t}{2}$
(2) **a** $2n$ **b** $5n + 6$ **c** $3n - 4$
(3) $5.2n$
(4) **a** $48x + 40$ **b** 7 tickets

Skills boost 4 Substituting into formulae

Guided practice

a $d = st$
   $d = 5 \times \underline{8}$
     $= \underline{40}$
b $d = st$
   $30 = s \times \underline{2}$
   $30 = \underline{2} s$
     $s = \underline{15}$

(1) **a** 2   **b** 9   **c** 8   **d** 2
   e 9   **f** 1   **g** 5   **h** 3
(2) **a** $F = 15$ **b** $F = 3$
(3) **a** $P = 9$ **b** $P = 3.5$
(4) **a** $y = 12$ **b** $y = 8$
(5) **a** $d = 150$ **b** $u = 10$

Practise the methods
(1) **a** 20   **b** 12   **c** 45   **d** 28
(2) **a** $2n$ **b** $5n + 3$
(3) **a** $y = 7$ **b** $x = 9$
(4) **a** $80 + 5x$ **b** 7
(5) **a** $a = 2$ **b** $x = 10$
(6) **a** $m = 5$ **b** $x = 7$
(7) $s = 66$

Problem-solve!
(1) $C = 6.5x$
(2) $a = 6.4$
(3) **a** $x = 3$ **b** $y = 2$ **c** $x = -1$
   d $x = -8$ **e** $x = \frac{1}{2}$ **f** $x = \frac{1}{5}$
(4) $x = 6$ cm
(5) 6 cm ($x = 9$)
(6) $x = 5$
(7) **a** total charge   **b** number of hours worked
   c £130

Unit 3 Graphs

AO1 Fluency check

(1) **a** 3, 6   **b** 10 am, 10.30 am   **c** 20, 60
(2) **a** 60   **b** 30   **c** 15
(3) **a** 1.5   **b** 0.5   **c** 0.25
(4) **a** 25 km/h   **b** 20 km/h

(5) **Number sense**

   a 16   **b** 2   **c** 4   **d** 12

Confidence questions

(1) 4 km

(2) $\frac{1}{2}$ hour

(3) 12 km/h

Skills boost 1 Using coordinates

Guided practice

A = (3, 2)

B = (− **4** , 5)

C = (− **4** , − **3**)

D = (**1** , −4)

(1) E (3, 1)   F (2, 5)   G (−3, 4)

   H (3, −3)   I (−4, −2)   J (−2, 2)

   K (0, 4)   L (1, −2)   M (−2, 0)

(2)

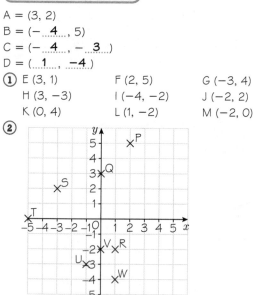

Skills boost 2 Drawing and interpreting distance–time graphs

Guided practice

a 3 pm

b 3 pm to **3.30** pm = **30** minutes

c **12.5** km in 1 hour = **12.5** km/h

d Speed = $\frac{\text{distance}}{\text{time}}$

      = $\frac{5}{2}$

      = **2.5** km/h.

(1) **a** 20 minutes

   b i 90 minutes   **ii** $1\frac{1}{2}$ hours

   c 5 km

   d 3.33 km/h

(2)

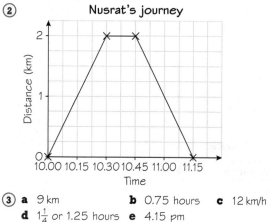

Nusrat's journey

(3) **a** 9 km   **b** 0.75 hours   **c** 12 km/h

   d $1\frac{1}{4}$ or 1.25 hours   **e** 4.15 pm

(4) **a**

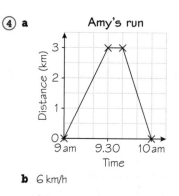

Amy's run

b 6 km/h

Skills boost 3 Drawing and interpreting real-life graphs

Guided practice

a Starting depth = 0 m

b Time to fill to 1 m = $\frac{3}{4}$ hours or 45 minutes

c Final depth = **2** m

(1) **a**

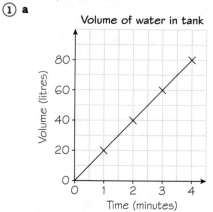

Volume of water in tank

b 50 litres

(2) **a** £20   **b** £30

(3) **a** £70   **b i** £30   **ii** £20

Practise the methods

(1) **a** £2   **b** £4   **c** £5   **d** £1

(2) **a** A (4, 0), B (3, −2), C (−4, −5), D (−2, 3)

   b

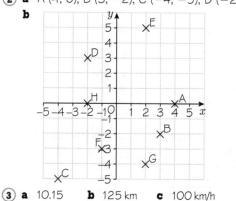

(3) **a** 10.15   **b** 125 km   **c** 100 km/h

① **a** 92 cm   **b** 57–58 minutes

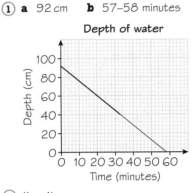

Depth of water

② (1, −1)

③ **a** Cylinder B
   b Cylinder A line 2, Cylinder B line 1

④ Campers R Us 14 days, C = £1600;
   Ready Camp 14 days, C = £2100.
   Campers R Us is cheaper. (Students may do
   calculation or add line to graph.)

Unit 4 Sequences

| AO1 Fluency check |

① **a** 10   **b** 5   **c** 13
② **a** 1   **b** 4   **c** 9   **d** 16
③ **a** 2^3 **b** 3^2 **c** 5^4
④ **a** 3   **b** 2   **c** 7

⑤ Number sense

   a −8, −6, −2, 6, 8
   b 0, 5, 25, 30, 35

Confidence questions

① **a**

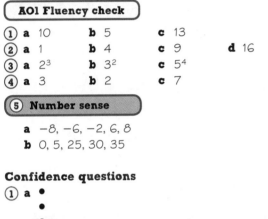

   b 11
② **a** 13   **b** 5
③ 11, 14, 17, 20

Skills boost 1 Generating sequences from patterns

| Guided practice |

a

b 2   4   6   8   **10**   **12**   14
   The 7th pattern has **14** tiles.

① **a**

   b 1, 4, 9, 16, 25
   c $1^2 = 1$, $2^2 = 4$, $3^2 = 9$, $4^2 = 16$, $5^2 = 25$

② **a**

   b 1, 3, 6, 10, 15, 21
         +2 +3 +4 +5 +6

③ **a**

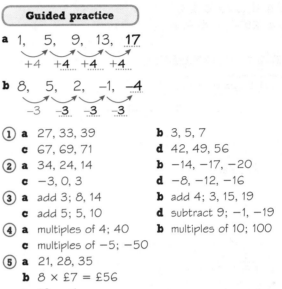

   b 18

Skills boost 2 Finding terms in a sequence

| Guided practice |

a 1,   5,   9,   13,   **17**
      +4   +4   +4   +4

b 8,   5,   2,   −1,   **−4**
      −3   −3   −3   −3

① **a** 27, 33, 39   **b** 3, 5, 7
   c 67, 69, 71   **d** 42, 49, 56
② **a** 34, 24, 14   **b** −14, −17, −20
   c −3, 0, 3   **d** −8, −12, −16
③ **a** add 3; 8, 14   **b** add 4; 3, 15, 19
   c add 5; 5, 10   **d** subtract 9; −1, −19
④ **a** multiples of 4; 40   **b** multiples of 10; 100
   c multiples of −5; −50
⑤ **a** 21, 28, 35
   b 8 × £7 = £56
   c 12 weeks

Skills boost 3 Using term-to-term rules

| Guided practice |

a 16,   11,   6,   1
         −5   −5   −5

b 16, 11, 6, 1, −4, **−9**, **−14**
① **a** 3, 10, 17, 24   **b** 18, 16, 14, 12
   c 20, 12, 4, −4   **d** −3, 1, 5, 9
   e −1, −3, −5, −7   **f** 20, 45, 70, 95
② **a** 3, 5, 7, 9   **b** 16, 13, 10, 7
   c 13, 11, 9, 7   **d** −5, −2, 1, 4
③ **a** 30   **b** −4
④ 8, 13

Practise the methods

① **a**

   b 3, 5, 7, 9, 11
   c 13
   d No, there are 13 sticks
② **a** 6   **b** −9
③ **a** 26   **b** 45
④ **a**   3 × 3   4 × 4   5 × 5   6 × 6
            9   16   25   36
   b square numbers
⑤ 48
⑥ 10, 7, 4, 1, −2
⑦ £33

Problem-solve!

(1) a 4   **b** $\frac{3}{4}$

(2) a 1st term 4, rule 'add 3'
   b 1st term 12, rule 'subtract 6'
   c 1st term -8, rule 'add 3'
   d 1st term -2, rule 'subtract 5'

(3) -5

(4) No, all terms are odd.

(5) 1, 2, 3

(6) A ii, B iii, C i, D v, E iv

(7) a 28   **b** 6

Unit 5 Triangles

A01 Fluency check

(1) a   **b**

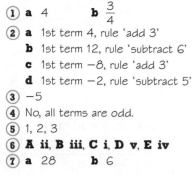

  c

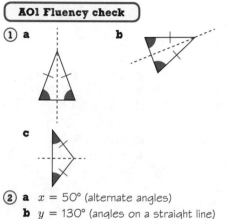

(2) a $x = 50°$ (alternate angles)
   b $y = 130°$ (angles on a straight line)

(3) a $x = 70°$ **b** $x = 49°$ **c** $x = 60°$

Confidence questions

(1) $x = 37°$ **(2)** $y = 59°$ **(3)** $z = 145°$

Skills boost 1 Finding angles in triangles

Guided practice

$180 - 90 - \underline{33} = \underline{57}$
$x = 57°$
Angles in a triangle add up to $\underline{180}°$.

(1) a 180°   **b** 180°   **c** 180°

(2) a 60°   **b** 65°   **c** 70°

(3) a $x = 50°$ **b** $y = 60°$

(4) a

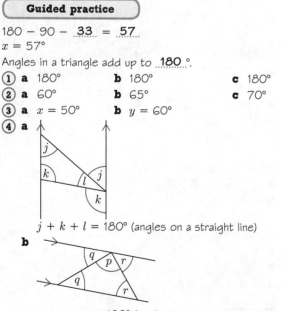

$j + k + l = 180°$ (angles on a straight line)

  b

$p + q + r = 180°$ (angles on a straight line)

(5) 55°

Skills boost 2 Finding sides and angles in triangles

Guided practice

2 equal sides → $\underline{\text{isosceles}}$ triangle.
$180 - \underline{35} - \underline{35} = \underline{110}$
          $x = 110°$
(angles in a triangle add up to 180°)

(1) a 65°, 6 cm   **b** 54°
   c 9 cm, 36°

(2) a $p = 55°$, $q = 70°$ **b** $r = 20°$, $s = 140°$

(3) a equilateral
   b **i** $3x = 180°$
      ii $AC = BC = 5$ cm

(4) 42°

Skills boost 3 Using exterior angles in triangles

Guided practice

$x = 50° + \underline{60}°$
$x = 110°$
Exterior angle equals sum of opposite interior angles.
$y = \underline{180}° - 110°$
$y = 70°$
Angles on a $\underline{\text{straight}}$ $\underline{\text{line}}$.

(1) a   **b**

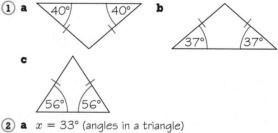

(2) a $t = 100°$ (exterior angle equals sum of opposite interior angles)
   b $v = 60°$ (exterior angle equals sum of opposite interior angles)

(3) 180°, 180°, 180°

(4) $\angle ACD = 65°$ (exterior angle equals sum of opposite interior angles)

Practise the methods

(1) a   **b**

  c

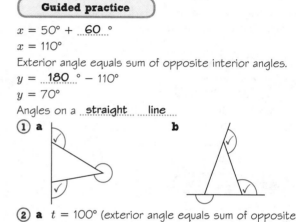

(2) a $x = 33°$ (angles in a triangle)
   b $y = 60°$ (angles in equilateral triangle)

(3) a $x = 62°$ (base angles of isosceles triangle)
      $y = 56°$ (angles in a triangle add up to 180°)
      $z = 7$ cm (equal side in isosceles triangle)
   b $p = 44°$ (angles in a triangle add to 180°)
      $q = 10$ cm (equal side in isosceles triangle)

(4) $x = 154°$ (angles on a straight line)
   $y = 108°$ (exterior angle equals sum of opposite interior angles)

(5) a $\angle ACQ = 117°$ (base angles in isosceles triangle and angles on a straight line)
   b $\angle ABP = 126°$ (exterior angle equals sum of opposite interior angles)

Problem-solve!

①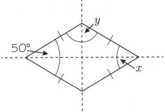

② 120°

③ **a** $x = 28°$ (angles on straight line add up to 180°)

   b $\angle ABC = 28°$ (base angles of isosceles triangle),
     $y = 124°$ (angles in triangle add up to 180°)

④ **a** $\angle ADB = \angle ABD = 60°$

   b equilateral

   c $\angle CDB = \angle DBC = 70°$

   d $\angle ADC = 130°$

⑤ **a**

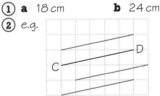

   b i $x = 50°$ **ii** $y = 130°$

⑥ **a** $\angle ZWV = 45°$ (half a right angle)

   b $\angle WZV = \angle WVZ$ (base angles of isosceles triangle)
     $= 67.5°$

   c $\angle VZY = 22.5°$ (angles in a right angle add up to 90°)

Unit 6 Quadrilaterals

AO1 Fluency check

① **a** 18 cm   **b** 24 cm

② e.g.

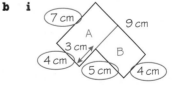

③ 16 squares

Confidence questions

① Perimeter 54 cm, area 126 cm²

② (2, 0)

③ 150°

Skills boost 1 Finding area and perimeter of shapes made from rectangles

Guided practice

a 8 cm + 10 cm + __3__ cm + 4 cm + 5 cm + __6__ cm
   = __36__ cm
   Perimeter = 36 cm

b For rectangle A   area = length × width
                        = 6 × __8__ = 48 cm²
   For rectangle B   area = length × width
                        = __4__ × 3 = 12 cm²
   Total area = area A + area B
               = __48__ + __12__ = 60 cm²

① 80 mm²

② **a** 72 cm²   **b** 26 cm²

③ **a** 6 cm, 5 cm   **b** 44 cm   **c** 90 cm²

④ **a** 58 mm   **b** 120 mm²

Skills boost 2 Finding coordinates of missing vertices

Guided practice

C = (2, __4__)

① H (2, 3)   ② L (−2, −3)

③ S (5, 2)   ④ Z (0, −3)

Skills boost 3 Finding angles in quadrilaterals

Guided practice

360 − 120 − __100__ − 90 = __50__
$x = 50°$
Angles in a quadrilateral add up to 360°.

① 360°, 360°

②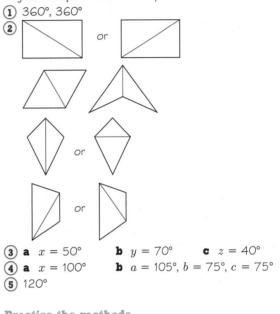

③ **a** $x = 50°$ **b** $y = 70°$ **c** $z = 40°$

④ **a** $x = 100°$ **b** $a = 105°$, $b = 75°$, $c = 75°$

⑤ 120°

Practise the methods

① **a i** 9 cm, 5 cm   **ii** 40 cm

   b i 7 mm, 14 mm   **ii** 52 mm

② **a i**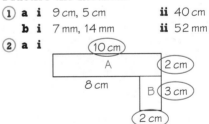

   ii A: 10 × 2 = 20, B: 3 × 2 = 6

   iii 26 cm²

   b i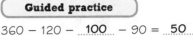

   ii A: 7 × 4 = 28, B: 5 × 4 = 20

   iii 48 cm²

③ (−4, 3)

④ 70°

⑤ $x = 65°$

Problem-solve!

① Perimeter 24 cm, area 36 cm²

② 575 cm²

③ Both 120°

④ Area 148 m², perimeter 62 m

⑤ **a** D (3, 4)   **b** E (1, −2)

⑥ 360 + 180 = 540°

Unit 7 Angles

① **a** $w = 60°$ **b** $x = 158°$
   c $y = 135°$ **d** $z = 58°$
② **a** **b**

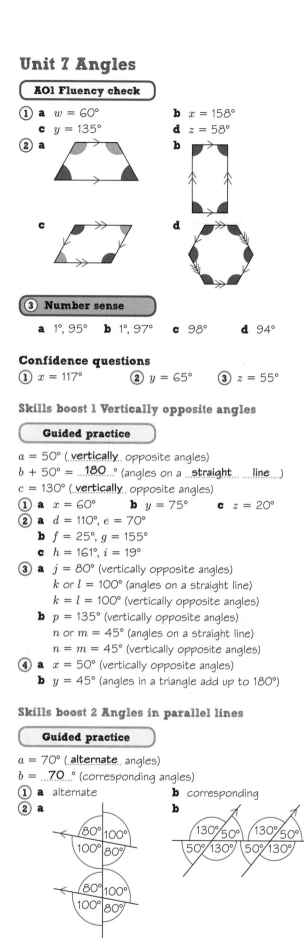

③ **Number sense**

   a 1°, 95° **b** 1°, 97° **c** 98° **d** 94°

Confidence questions

① $x = 117°$ ② $y = 65°$ ③ $z = 55°$

Skills boost 1 Vertically opposite angles

Guided practice

$a = 50°$ (**vertically** opposite angles)
$b + 50° = $ **180** ° (angles on a **straight line**)
$c = 130°$ (**vertically** opposite angles)
① **a** $x = 60°$ **b** $y = 75°$ **c** $z = 20°$
② **a** $d = 110°, e = 70°$
   b $f = 25°, g = 155°$
   c $h = 161°, i = 19°$
③ **a** $j = 80°$ (vertically opposite angles)
      k or $l = 100°$ (angles on a straight line)
      $k = l = 100°$ (vertically opposite angles)
   b $p = 135°$ (vertically opposite angles)
      n or $m = 45°$ (angles on a straight line)
      $n = m = 45°$ (vertically opposite angles)
④ **a** $x = 50°$ (vertically opposite angles)
   b $y = 45°$ (angles in a triangle add up to 180°)

Skills boost 2 Angles in parallel lines

Guided practice

$a = 70°$ (**alternate** angles)
$b = $ **70** ° (corresponding angles)
① **a** alternate **b** corresponding
② **a** **b**

③ $f = 105°$ (angles on a straight line),
   $g = 105°$ (corresponding angles)
④ $y = 50°$. Several possible methods, e.g.
   $\angle CBE = 50°$ (angles on a straight line) then
   $y = 50°$ (corresponding angles).

Skills boost 3 Finding angles in shapes

Guided practice

$\angle ABE = $ **45** ° (**corresponding** angles)
$x = 180° − 80° − $ **45** ° (angles in a triangle add up to 180°)
$x = 55°$
① **a** $y = 54°$ (corresponding angles)
   b $x = 50°$ (angles in a triangle)
② $a = 108°$ (alternate angles), $b = 72°$ (angles on a straight line), $c = 95°$ (alternate angles), $d = 85°$ (angles on a straight line)
③ $a = 115°, b = 65°, c = 115°, d = 115°, e = 65°$
④ Several possibilities, e.g.
   a $\angle ABG = 40°$ (base angles of isosceles triangle)
      $\angle AGB = 100°$ (angles in a triangle)
   b $\angle BGC = 40°$ (alternate angle to $\angle ABG$)

Practise the methods

① **a** alternate **b** neither **c** neither
   d corresponding **e** neither **f** corresponding
② $x = 65°$ (angles on a straight line)
   $y = 115°$ (vertically opposite angles)
③ $a = 82°$ (vertically opposite angles), $b = 82°$ (alternate angle to a, or corresponding angle), $c = 98°$ (angles on a straight line)
④ **a** $\angle CGF = 75°$ (corresponding angles)
      $\angle CGH = 105°$ (angles on a straight line)
   b $\angle DGF = 105°$ (vertically opposite angles)
⑤ **a** $\angle DBC = 55°$ (corresponding angles)
   b $\angle C = 35°$ (angles in a triangle)

Problem-solve!

① $a = 64°$ (vertically opposite angles),
   $b = 116°$ (angles on a straight line),
   $c = 66°$ (angles on a straight line or round a point)
② $x = 90°$
③ **a** 62° (base angles of isosceles triangle)
   b 56° (angles in a triangle add up to 180°)
   c 56° (vertically opposite angles)
④ $\angle ACB = 62°$ (vertically opposite angles), $\angle BAC = 44°$ (alternate angles), $\angle CED = 74°$ (angles in a triangle add up to 180°), $\angle ABC = 74°$ (alternate angles). Triangle ABC has angles 62°, 44°, 74° and Triangle DEC has angles 62°, 44°, 74°.
⑤

Unit 8 Loci

A01 Fluency check

① Accurately drawn circle with radius 2.5 cm
② Accurately drawn line of length 6 cm
③ 6 equal sides, 6 equal angles

Confidence questions

(1)

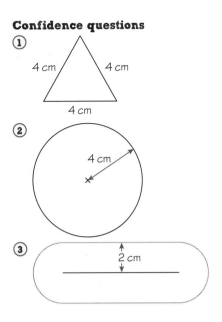

4 cm 4 cm

4 cm

(2)

4 cm

(3)

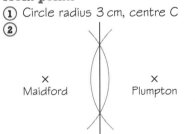

2 cm

Skills boost 1 Constructing triangles and hexagons

(1)

3 cm 3 cm

3 cm

(2)

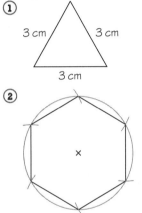

×

(3) Equilateral triangle, each side 6 cm ± 1 mm, construction arcs left in.

Skills boost 2 Constructing loci equidistant from points

(1) Circle radius 3 cm, centre C

(2)

×   ×
Maidford   Plumpton

(3) Shaded circle radius 4 cm, centre P

Skills boost 3 Constructing loci around lines and corners

(1)

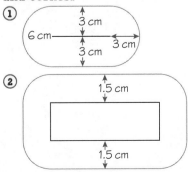

3 cm
6 cm
3 cm 3 cm

(2)

1.5 cm

1.5 cm

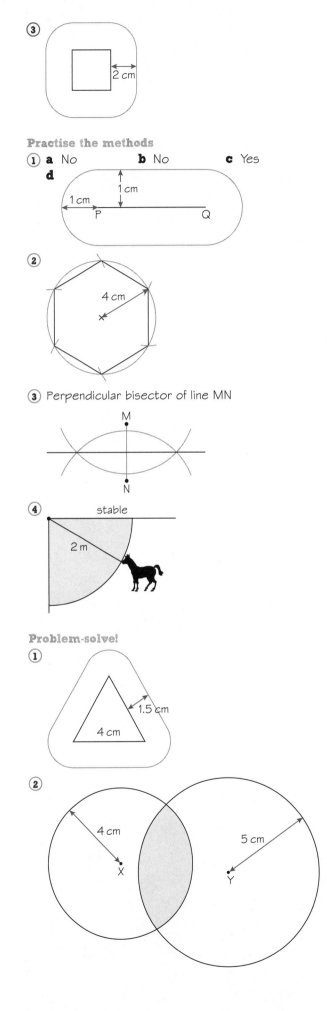

(3)

2 cm

Practise the methods

(1) **a** No **b** No **c** Yes

d

1 cm
1 cm
P   Q

(2)

4 cm

×

(3) Perpendicular bisector of line MN

M
•

•
N

(4)

stable

2 m

Problem-solve!

(1)

1.5 cm

4 cm

(2)

4 cm

X

5 cm

Y